1346

507.24 Franklin, Allan,
FRA   1938-

        The neglect of
           experiment

     $44.50

DATE			

The neglect of experiment

The neglect of experiment

Allan Franklin

Department of Physics
University of Colorado

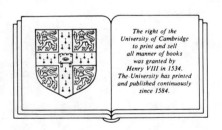

The right of the
University of Cambridge
to print and sell
all manner of books
was granted by
Henry VIII in 1534.
The University has printed
and published continuously
since 1584.

Cambridge University Press

1346

Cambridge

London   New York   New Rochelle

Melbourne   Sydney

Published by the Press Syndicate of the University of Cambridge
The Pitt Building, Trumpington Street, Cambridge CB2 1RP
32 East 57th Street, New York, NY 10022, USA
10 Stamford Road, Oakleigh, Melbourne 3166, Australia

First published 1986

Printed in the United States of America

Library of Congress Cataloging-in-Publication Data
Franklin, Allan, 1938–
The neglect of experiment.
Bibliography: p.
Includes index.
1. Science – Experiments – Philosophy.   2. Science –
Philosophy.   3. Physics – Philosophy.   I. Title.
Q182.3.F73   1986   507'.24   86–2604

British Library Cataloging-in-Publication Data
Franklin, Allan
The neglect of experiment.
1. Science – Philosophy
I. Title
501   Q175

IBSN 0 521 32016 X

To Jane, for lots of reasons

Contents

Acknowledgments	*page* ix
List of abbreviations	xiii
Introduction	1
1 The discovery of parity nonconservation	7
2 The nondiscovery of parity nonconservation	39
3 CP or not CP	73
4 The role of experiment	103
5 Do experiments tell us about the world?	138
6 The epistemology of experiment	165
7 The epistemology of experiment: case studies	192
8 Forging, cooking, trimming, and riding on the bandwagon: fraud in science	226
Conclusion	244
Notes	245
Index	281

Acknowledgments

Since changing research fields from high-energy physics to the history and philosophy of science, I have had the assistance and encouragement of a large number of people. One of my largest debts is to the members of the Department of History and Philosophy of Science, Chelsea College, University of London. Heinz Post and Michael Redhead, who have headed the department during my visits over the last eight years, along with Jon Dorling (now at the University of Amsterdam), Donald Gillies, and Moshe Machover, have been constant sources of knowledge, constructive criticism, and inspiration. At least part of the impetus for my work comes from the fact that each year I am asked to give a seminar in the department. I still do not have the courage to admit to them that I have nothing new to say. They have been not only colleagues but friends. I am proud to be a member of the department.

Thanks are also due to my friend, research collaborator, and all-around boon companion Colin Howson of the Philosophy Department, London School of Economics. Our collaboration has been an unadulterated pleasure for me, and it has been a true collaboration. It is, I believe, fair to say that our joint work, much of which is included in this book, could not have been done by either of us separately. Our discussions while walking across Hampstead Heath have given me a new conception of the term "peripatetic philosophy." I have learned much from him and look forward to our continued work together. I have also benefited greatly from discussions with other members of the Philosophy Department at LSE: Peter Urbach, John Watkins, John Worrall, and Elie Zahar.

Several years ago, an unsympathetic referee remarked of one of my papers that it was "typical of the kind of work done by the

University of London crowd." Although that was not intended as a compliment, I have always considered it one.

My friends and colleagues at the University of Colorado have also been helpful and supportive. Howard Smokler has always been available to provide knowledge of the philosophy of science, to discuss issues, and to encourage my work. Some of our collaborative work is also included here. Jim Scott and Jim Warwick, two distinguished scientists, have always been willing to discuss my work with me and to provide illustrations of various points from their own experience. I have had valuable discussions with members of the Physics Department, including David Bartlett, Asim Barut, Paul Beale, Peter Bender, Jinx Cooper, John Cumalat, Tom DeGrand, Joseph Dreitlein, William Ford, Mark Handschy, K. T. Manhanthappa, Uriel Nauenberg, Bill O'Sullivan, and Brian Ridley.

As the source citations in this book will show, I owe a large debt to Ian Hacking, not only for his encouragement and helpful conversations but also because his book *Representing and Intervening* has made it legitimate to discuss the philosophy of experiment. If this book is less philosophical than it should be, that is not the fault of Nancy Cartwright. Nancy has always tried, in her invariably constructive criticism, to help me to address issues as would a trained philosopher of science. I hope she has had some success, but it probably is not as much as she would have liked. The fine editorial hand of John Heilbron can be seen directly in the sections on CP violation and on Millikan, and indirectly everywhere else. I always try to write with his red pencil in mind. Our conversations on the history of twentieth-century physics and on the nature of the history of science have been very valuable.

Harvey Brown was the first person to point out to me the relevance of my work to the Duhem–Quine problem, and he was instrumental in obtaining an invitation for me to give a series of seminars on experiment in physics at the University of Campinas, Brazil, in 1982, on which part of this book is based. William Fairbank, Jr., has graciously allowed me to use our joint work on Millikan here.

Many of the participants in the episodes recounted here have provided both valuable discussions and materials. T. D. Lee gave me his unpublished history of the weak interactions. My correspondence with Richard Cox gave me valuable insights into the early experiments showing parity violation. Val Fitch and Jim Cronin gave me preprints of their Nobel lectures and copies of their original proposal. Fitch also provided a copy of their laboratory notebook, which was invaluable in analyzing their experiment. Jim Christenson also provided a copy of the proposal. Materials from the Millikan notebooks, as well as valuable discussions, were provided by Judy Goodstein and are published courtesy of the California Institute of Technology Archives. Mircea Fotino first pointed out the relevance of the work of Keith Porter to my question on the epistemology of experiments. Both he and Professor Porter provided materials and useful discussions.

At various times I have had the benefit of conversations with Sir Karl Popper, Noretta Koertge, Peter Galison, Andy Pickering, and David Gooding.

This book would not exist were it not for the tremendous efforts of Mrs. Evelyn Jones, who typed the manuscript. She has treated this work as if it were her own, and she has my gratitude. Other members of the Physics Department staff, Joan Sonnenberg, Linda Frueh, Margaret Kneebone, and Kathy Oliver, also helped with the manuscript.

Portions of this work were supported by the National Science Foundation under grants SES-8204074 and SES-8308260. Any opinions, findings, conclusions, or recommendations expressed in this publication are those of the author and do not necessarily reflect the views of the National Science Foundation. I also received support from the Graduate School, the Council on Research and Creative Work, the Committee on University Scholarly Publications, and the Physics Department of the University of Colorado.

Sections of this book were originally published elsewhere. Chapters 1 and 2 appeared in "The Discovery and Nondiscovery of Parity Nonconservation," *Studies in History and Philosophy of Science*, *10* (1979), 201–57. Portions of Chapter 1 also were contained in "Justification of a 'Crucial' Experiment: Parity Nonconservation," with Howard Smokler, *American Journal of Physics*, *49* (1981), 109–12. Chapter 3 appeared in "The Discovery and Acceptance of CP Violation," *Historical Studies in the Physical Sciences*, *13(2)* (1983), 207–38. Portions of Chapter 4 were published as "Are Paradigms Incommensurable?" *British Journal for the Philosophy of Science*, *35* (1984), 57–60; "Why Do Sci-

xii   *Acknowledgments*

entists Prefer to Vary Their Experiments?" with Colin Howson, *Studies in History and Philosophy of Science*, *15* (1984), 51–62; "Newton and Kepler, A Bayesian Approach," with Colin Howson, *Studies in History and Philosophy of Science*, *16* (1985), 379–85; and "A Bayesian Analysis of Excess Content and the Localisation of Support," with Colin Howson, *British Journal for the Philosophy of Science*, *36* (1985), 425–31. Portions of Chapter 5 appeared in "Millikan's Published and Unpublished Data on Oil Drops," *Historical Studies in the Physical Sciences*, *11* (1981), 185–201; and "Did Millikan Observe Fractional Charges on Oil Drops?" with William Fairbank, Jr., *American Journal of Physics*, *50* (1982), 394–7. A preliminary sketch of Chapter 6 appeared in "The Epistemology of Experiment," *British Journal for the Philosophy of Science*, *35* (1984), 381–90. Chapter 8 was contained in "Forging, Cooking, Trimming, and Riding on the Bandwagon," *American Journal of Physics*, *52* (1984), 786–93. I am grateful to Pergamon Press, the University of California Press, the British Society for the Philosophy of Science, and the American Association of Physics Teachers, who hold the copyrights, for permission to reproduce the materials in this book.

Abbreviations

AJP	*American Journal of Physics*
CR	Académie des Sciences, Paris, *Comptes rendus hebdomadaires des séances*
JETP	*Zhurnal Eksperimental'noi i Teoreticheskoi Fiziki*
Natur.	*Naturwissenschaften*
NC	*Il Nuovo Cimento*
Phil. Mag.	*Philosophical Magazine*
PR	*Physical Review*
PRL	*Physical Review Letters*
PL	*Physics Letters*
Phys. Z.	*Physikalische Zeitschrift*
Proc. Roy. Soc. (London)	*Proceedings of the Royal Society (London)*
Z. Phys.	*Zeitschrift für Physik*

Introduction

One of the great anticlimaxes in all of literature occurs at the end of Shakespeare's *Hamlet*. On a stage strewn with noble and heroic corpses – Hamlet, Laertes, Claudius, and Gertrude – the ambassadors from England arrive and announce that "Rosencrantz and Guildenstern are dead." No one cares. A similar reaction might be produced among a group of physicists,[1] or even among historians and philosophers of science, were someone to announce that "Lummer and Pringsheim are dead." And yet they performed some of the most important experiments in the history of modern physics. It was their work on the spectrum of black-body radiation,[2] along with that of Rubens and Kurlbaum,[3] that showed deviations from Wien's Law and formed an important part of the background to Planck's introduction of quantization.

This is symptomatic of the general neglect of experiment and the dominance of theory in the literature on the history and philosophy of science. In Thomas Kuhn's history of quantization, *Black-Body Theory and the Quantum Discontinuity, 1894–1912*,[4] Lummer, Pringsheim, Rubens, and Kurlbaum are, at best, peripheral characters. The title indicates what Kuhn thinks is important. We never see what the experimental results were or find a discussion of how they were obtained.

But, it might be said, that is only an isolated case. Surely everyone is aware of the famous experiments of Galileo and the Leaning Tower of Pisa, of Thomas Young's double-slit interference experiment, and of the Michelson–Morley experiment. What seems to be generally known, particularly by scientists, about these experiments shows the mythic treatment of experiment. Real experiments and their roles are not often dealt with.

According to the myth, Galileo dropped two unequal weights from the top of the tower, observed that they fell at equal rates, and thereby refuted Aristotelian mechanics and established the importance of experiment in physics. There are several problems with this story. First, there is serious doubt that Galileo ever performed the experiment.[5] Had he done so, he would have gotten rather confusing results. A modern replication of the experiment showed that in a fall of 200 feet, a shotput will beat a softball by 20 to 30 feet.[6] This is not a large enough difference to satisfy the Aristotelian theory, but it is too large for equality of fall, or the hand's-breadth difference Galileo reported. Even had Galileo done the experiment and observed his reported results, an Aristotelian could easily have modified the theory to accommodate the data.[7]

In the case of Young's experiments, John Worrall has argued that, contrary to popular belief, these experiments did not decisively refute the corpuscular theory of light and establish the wave theory.[8] He points out that corpuscular explanations of both interference and diffraction were available. In addition, until Fresnel's later work, the wave theory could not explain the rectilinear propagation of light, which was at least as serious a problem for it as interference was for the corpuscular model.

Similarly, the Michelson–Morley experiment was supposed to have demonstrated the nonexistence of the ether and to lead directly to Einstein's special theory of relativity. Although Einstein's 1905 paper on relativity did mention the failure of the then recent attempts to measure the velocity of the earth relative to the ether, no specific mention was made of the work of Michelson and Morley. Historians of science have disagreed on the importance of this work for Einstein's theory, but in no case do they assert the kind of importance it has been given in more popular accounts such as textbooks. This error can, of course, be attributed to physicists' lack of knowledge of the history of their discipline, and this is partially true, but similar accounts appear in philosophical discussions.[9] Even a detailed historical study of the Michelson–Morley experiment[10] failed to point out that in their 1887 paper there was an important difference between the raw data given in the tables, which showed a large linear drift, and the graph of residuals, which gave the well-known null result. Actually, Michelson and Morley set an upper limit for the velocity of the earth relative to the ether: one-sixth of the earth's orbital velocity. An excellent study of this has been published by Mark

Handschy,[11] who has given a plausible reconstruction of the method used to subtract the drift. The point here is that real, as opposed to mythological, experiments are rarely discussed, even when experiment is mentioned at all.

Fortunately, recent work on actual experiments seems to be reversing this trend, or at least modifying it. On the philosophical side there is Ian Hacking's excellent book *Representing and Intervening*,[12] which argues persuasively, using numerous examples and illustrations from the practice of science, against the idea of theory-dominated experiment and in favor of the view that experiment often has a life of its own. Historians of science such as Peter Galison, Andrew Pickering, David Gooding, John Worrall, Bruce Wheaton, and Roger Stuewer have presented detailed accounts of various experimental episodes.[13]

This book is intended as a contribution to this continuing study of the history and philosophy of experiment. It will deal primarily with two questions.

1. What role does, and should, experiment play in the choice between competing theories or hypotheses or in the confirmation and support of theories or hypotheses?
2. How do we come to believe rationally in the results of an experiment, or how do we separate a result, obtained by use of an apparatus to measure or observe a quantity, from an artifact created by the experimental apparatus?

In answering the first question, philosophers of science, with the exceptions of Popper, Glymour, Hacking, and Shapere,[14] seem either to undervalue the role of experiment, where they acknowledge it at all, or to take observational or experimental results as given and unproblematical. The philosophical positions range from what one might call the sociological and psychological views of Feyerabend and Kuhn to the logical problems raised by Duhem and Quine.

In *Against Method*,[15] Feyerabend argues for methodological anarchy in science. He attributes scientific change to propaganda victories by one group of scientists over another and allows no role for experiment in determining the decision between two competing theories. Kuhn, in *The Structure of Scientific Revolutions*,[16] argues that major scientific change occurs by paradigm shift. He regards two competing paradigms as incommensurable, and although he does allow for rational discussion and experimental evidence as parts of the decision-making process, he does not seem to regard them as major components. He states that "The competition between paradigms is not the sort of battle that can be resolved by proofs."[17] In Kuhn's view, there can be no falsifying instances or crucial experiments.

A similar position is taken by Lakatos, who argues that a theory can be rejected on the basis of experimental evidence only if an alternative and better theory is available: "... no experiment, experimental report, observation statement or well-corroborated low level falsifying hypothesis alone can lead to falsification. There is no falsification before the emergence of a better theory."[18]

Duhem and Quine[19] have raised a logical objection to the role that experiment plays in the refutation of hypotheses. They have argued that any theory or hypothesis can be saved from refutation by some suitable adjustment in background knowledge or by auxiliary hypotheses. Quine states that "any statement can be held true come what may, if we make drastic enough adjustments elsewhere in the system."[20] If this is so, then only the whole of science can be affected by experimental evidence.

Recent work on the confirmation of scientific theories or hypotheses tends to regard observations or experimental results as given,[21] although Glymour[22] does offer historical examples to support his bootstrap model of confirmation.

The second question regarding the epistemology of experiment has been almost totally neglected by philosophers of science, except for the recent work of Ian Hacking and Dudley Shapere referred to earlier.

The approach taken in this study will be to combine detailed historical study of episodes in the history of twentieth-century physics with a discussion of some of the philosophical issues. These episodes will include (1) the experiments on parity nonconservation, both in 1957, when the violation was discovered, and in 1930, when it could have been but wasn't, (2) the discovery and acceptance of CP violation, and (3) Millikan's oil-drop experiments, which established the quantization of electric charge and also measured e, the fundamental unit of charge. As I admit in Chapter 4, I do not have a general answer to the first question concerning the role of experiment in theory choice. These episodes do, however, provide examples of "crucial" and "convincing" experiments that played major roles in theory choice. I shall also argue that this role can be philosophically justified in these particular episodes. This provides at least some counterexamples to those philosophers and sociologists of science, some of them discussed briefly earlier, who would deny that role.

That role demands that we have good reasons for believing in experimental results. The second half of this book is devoted to discussing how we can come to believe rationally in these results. This, too, has been questioned. In Chapter 6 I shall present a set of strategies, used by practicing scientists (and which I argue can be independently justified), that provide us with rational grounds for belief. Here, too, the emphasis will be on the actual practice of science, and the three episodes mentioned earlier will provide some of the evidence. I shall also discuss two other questions concerning experiment: the possible undue influence of theoretical preconceptions, and the question of fraud in science.

This study is certainly not a complete discussion of experiment in physics. I have concentrated on only one of the many roles that experiment plays: that involved in theory choice or the confirmation of theories. In Chapter 4 we shall discuss some of the other roles it plays.

A few words here on the methodology I have used: I work primarily with what philosophers of science have called the context of justification, and consequently with published papers, the artifacts that scientists have submitted for peer review and as their contributions to the permanent record. This is not to say that publication is the sole means of communication among scientists. Preprint circulation, correspondence, conversation, and attendance at meetings and conferences are also important methods of information exchange. Publications also do not give a complete picture of an experiment. Laboratory notebooks, correspondence, and interviews give us more information, and the laboratory notebooks will be used in two of the episodes to be discussed. I do believe, however, that for the questions this study will consider–theory choice or confirmation and the validation of experimental results – the published record should be used. I think that the information acquired by an experimenter, by any means, is essentially that contained in the published work, and I think that the published reasons given both for the motivation of the experiment and for the acceptance or rejection of hypotheses are those that in fact determined the course of the work. Whatever an experimenter's private reasons for believing in a result, I think that only those that the author is willing to state publicly should be considered in discussing the validity of those results.

1

The discovery of parity nonconservation

In this chapter we shall examine the history of the 1957 discovery of parity nonconservation, or the violation of mirror symmetry in nature. The discovery began with the suggestion that parity was not conserved, a suggestion put forth as a means of solving a particularly vexing problem that had resisted all conventional attempts at solution. The revolutionary nature of this suggestion was clearly recognized by the authors, and that led them to suggest other experiments that would test the principle. Those experiments were then performed, with positive results, and that led to the overthrow of the principle of mirror symmetry. They were thus "crucial" experiments, in the sense that they unambiguously, and within a short period of time, decided the issue between two theories (or classes of theories) of some importance for a certain segment of the scientific community. We shall examine the evidence for this in Section 1.1.

What was not realized at that time was that there were already two experimental results in the literature, published in 1928 and 1930, that had the same physical and logical content and showed parity violating effects, but whose implications and significance were not recognized either by the authors or by any member of the physics community until after 1957.

It seems clear, then, that it is not only the physical and logical content of an experiment or series of experiments that determines its crucial nature; other conditions must also be present. Otherwise, one could not have the case of an experiment that is crucial at one time but not at another time. Although one would not wish to overgeneralize from a single case study, there are several points worth noting. The first, and most obvious, is that the appropriate scientific community must recognize the relevance of the experimental results for deciding the issue between two competing theories. In fact, in this case, there was no theory competing with the strongly held principle of parity conservation. The requirement of an alternative theory is not, however, absolute, because it was recognized that the experimental results violated mirror symmetry without such an alternative. The second point is that even when an anomalous result (although not its implications) is recognized, as it was in this case, the anomaly must appear to be of sufficient interest and importance to stimulate further experimental and theoretical work. This will be made clear in Chapter 2, where the experiments of 1928 and 1930 are discussed in detail and it is argued that they do demonstrate parity nonconservation, but were neglected. Later in this chapter we shall also examine the intellectual context prior to 1957 and discuss why such a revolutionary suggestion as parity nonconservation was taken seriously.

1.1 THE 1957 DISCOVERY

Perhaps the clearest case of a crucial experiment or set of crucial experiments in the history of physics occurred with the discovery of nonconservation of parity in 1956 and 1957. We can briefly summarize the history of that discovery as follows: The existence of an anomaly in the decay of certain elementary particles, the so-called θ–τ puzzle, led T. D. Lee and C. N. Yang[1] to propose that the well-established principle of parity conservation, or reflection symmetry, was violated in the weak interactions.[2] Their paper also proposed several experiments that would test their suggestion. Their hypothesis was corroborated in a set of experiments performed by Wu and associates,[3] by Garwin and associates,[4] and by Friedman and Telegdi.[5] Subsequently, other experiments corroborated their proposal in several other situations. Before discussing the detailed history of this discovery, we shall discuss what is meant by the concept "parity."

The parity operation is defined as the reflection of all spatial coordinates through the origin of the coordinate system (i.e., $x \rightarrow -x, y \rightarrow -y, z \rightarrow -z$).[6] If all the physical laws remain invariant under this operation, we speak of them as having reflection, or space inversion, symmetry. In this case we also say that parity is conserved.[7] As shown in Figure 1.1, the parity operation changes a right-handed coordinate system into a left-handed one, and vice versa. Thus, invariance under space inversion, or parity conservation, is equivalent to indistinguishability of left and right.

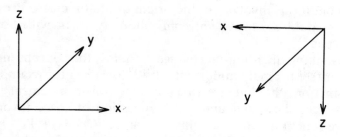

RIGHT HANDED
COORDINATE SYSTEM

LEFT HANDED
COORDINATE SYSTEM

Figure 1.1   Right-handed and left-handed coordinate systems.

Although this symmetry can be applied to classical physical systems, its most important applications occur in quantum-mechanical systems. The indistinguishability of left and right occurs in classical discussions of the interaction of electric currents and magnetic fields. All students of introductory physics are taught that given the direction of an electric current, one finds the direction of the magnetic field caused by that current by applying a "right-hand rule."[8] If one wishes to find the direction of the force exerted by the field on another current, one applies a second right-hand rule. The handedness of these rules is, however, arbitrary. If one deals only with observable quantities, the currents and the force, one finds that two left-hand rules will give an identical result, showing the symmetry between left and right.

Gibson and Pollard point out that similar questions arise in discussions of chemical and biological processes:

For example, certain complex molecules exist in two forms: isomers having the same formula, but whose structures are related to each other by the operation of space inversion. [See Figure 1.1.] Solutions of these substances which contain an excess of one type of molecule are optically active, and the two isomers can thus be distinguished as dextro- or laevo-rotary (D or L).[9] The amino acids of living matter are of this type and it is found that only L amino acids occur in living systems, while in laboratory syntheses D and L types occur in equal amounts.[10]

Thus, although there does exist a left–right asymmetry in living matter, this is not a fundamental asymmetry involving the laws of nature, as indicated by the fact that laboratory work produces both types in equal amounts. It is rather a comment on the particular conditions involved in the origin of life on earth. Presumably, life could exist involving only the D type, or with both D and L types.[11]

In describing quantum-mechanical systems, the concept of parity becomes extremely important. If we consider, for example, a wave function, Ψ, that describes an electron in a central force field, we can show that under the operation of space inversion the wave function either remains the same, $\Psi(r) = +\Psi(-r)$, or changes sign, $\Psi(r) = -\Psi(-r)$. The latter state is said to have odd or negative parity, whereas the former state has even or positive parity. This spatial parity for a wave function of angular momentum, ℓ, is given by $(-1)^{\ell}$. One should note, however, that it is only the square of the total wave function, $|\Psi|^2$, which is composed of individual-particle wave functions, or the relative phases of several individual wave functions that are observable. These are not affected by a change in the sign of the individual-particle wave functions if the system contains only one type of particle. One might, however, have a system that contains several different kinds of particles and in which the properties of the individual-particle wave functions for different kinds of particles are different under space inversion. We refer to the behavior of each type of particle wave function as the intrinsic parity of that particle. The total parity of the system then consists of the product of the spatial parity and the intrinsic parity. Under these circumstances, a change in the sign of the particle wave function under inversion becomes meaningful. If the parity of the system changes by the destruction or creation of a particle of a given intrinsic parity, one can, in fact, observe the effect of this change. If the total parity of the system is conserved, then the spatial parity of the residual system must change in order to compensate for the change in intrinsic parity caused by the creation or destruction of the particle.[12] We shall consider several examples of this later. It should also be added that one finds experimentally that the intrinsic parity of the π meson or pion is odd or negative.[13] This will be very important in subsequent discussions.

The modern history of the study of parity conservation began in 1924 with the work of Otto Laporte.[14] In studying the spectrum of radiation emitted by iron atoms, Laporte found that he could classify the states of the iron atom into two types, which he called *gestrichene* and *ungestrichene*, or stroked and unstroked. He discovered that radiation was emitted only when the atom went from a state of one type to a state of the other (e.g., from a stroked state to an unstroked state). He was not able, at that time, to offer an explanation for the truth of this selection rule, and it has passed into the physics literature as Laporte's Rule.

In 1927, Eugene Wigner[15] gave an explanation of this rule based on the concept of parity conservation, or reflection symmetry. We can understand Wigner's argument as follows: The states that Laporte called stroked and unstroked are in fact states of positive parity and negative parity. Because the intrinsic parity of the emitted photon, in the most important case of electric-dipole radiation, is negative, then in order for the total parity of the system to be conserved, the parity of the atomic state must change. Thus, we get radiation emitted only when the atom changes from a state of positive parity to one of negative parity, or vice versa. Transitions between states of the same parity are forbidden by parity conservation. This concept of parity conservation quickly became an established principle of physics.[16] As Frauenfelder and Henley state, "Since invariance under space reflection is intuitively so appealing (why should a left- and a right-handed system be different?), conservation of parity quickly became a sacred cow."[17] We shall present evidence to support this view later in this section, but for the moment we shall assume that, following Wigner's paper, virtually all physicists accepted parity conservation.

That was the situation in the physics community until 1956.[18] At that time, one of the more puzzling problems facing physicists was the θ–τ puzzle. Yang summed up the problem in late 1956:

I now turn to the last topic, namely the question of the identity of the K particles. Interest in this subject started when it was pointed out that by studying the decay of the $K_{\pi 3}^+$, also known as the τ^+ meson, into three charged π mesons, it is possible to conclude something about its spin and parity. Let me give you a simple example of the kind of arguments involved. Suppose one of the π mesons resulting from the decay of a τ meson is very slow, so that one may assume that it comes out in an s state relative to the center of mass of the other two π mesons. What can one conclude under these circumstances? Well, the remaining two pions will be in one of the states $0^+, 1^-, 2^+, \ldots$, because the pions are spinless and their intrinsic parities add up to 'even' so that the spin and parity of the two pion system are identical with those of the orbital part of their wave function.[19] The extra π meson is odd, and contributes no angular momentum; we can therefore conclude that the spin and parity of the 3 pions together can only be 0^-, 1^+, 2^- ... *Since spin and parity are conserved in any process* [emphasis added], the τ meson spin and parity also must be 0^-, 1^+, 2^-, etc. The $K_{\pi 2}$ by the same argument can only be 0^+, 1^-, 2^+.... Therefore, we may conclude that if ever the τ meson is found to decay into three pions, one of which is slow, the $K_{\pi 2}$ and τ cannot be the same particle. Of course, the analysis is not quite so simple, because there is always the question: How slow is "slow" for the decay pion: To investigate this problem the distribution of the momenta of the decay products of the τ has been plotted in a triangular plot and the full distribution studied against the various possible spin parity assignments for the τ. With the one thousand or so points available at the present moment the general opinion is that the τ is not the same particle as the $K_{\pi 2} \equiv \theta^+$.

However, it will not do to jump to hasty conclusions. This is because experimentally the K mesons seem all to have the same masses and the same lifetimes. The masses are known to an accuracy of say from 2 to 10 electron masses or a fraction of a percent, and the lifetimes are known to an accuracy of say 20%. Since particles which have different spin and parity values, and which have strong interactions with the nucleons and pions, are not expected to have identical masses and lifetimes, one is forced to keep the question open whether the inference mentioned above that the τ^+ and θ^+ are not the same particle is conclusive. Parenthetically I might add that the inference would certainly have been regarded as conclusive, and in fact more well founded than many inferences in physics, had it not been for the anomaly of mass and lifetime degeneracies.[20]

A similar view, that gives some further insight into the nature of the problem was expressed by Lee in his recollections in 1971:

Among these new processes, the most puzzling ones were the charged θ and τ mesons. Both mesons were defined according to their decay modes:

$$\theta^+ \rightarrow \pi^+ \pi^0$$
and $$\tau^+ \rightarrow \pi^+ \pi^+ \pi^-$$
or $$\tau'^{\,+} \rightarrow \pi^+ \pi^0 \pi^0.$$

The spin-parity of the θ^+ is clearly 0^+, 1^-, 2^+, etc. As early as 1954, Dalitz[21] already pointed out that the spin-parity of the τ^+ can be analyzed through his Dalitz plot, and the existing data were more consistent with the assignment 0^- than 1^+. Although both mesons were known to have comparable masses (within 20 MeV), there was nothing extraordinary about it. However, by 1955, very accurate life-time measurements became available. This, together with a statistically much more significant Dalitz plot of τ^+ decay, presented a very puzzling picture indeed. Except for very high spin assignments $J \geq 3$ (which was not preferred partly on esthetic grounds and partly because it would lead to angular correlations between the production and decay directions, while none were observed), the spin-parity of the τ^+ was determined to be definitely 0^-; therefore, it appeared to be definitely a different particle from θ^+. Yet within the experimental accuracy ($\sim$ a few percent) these two particles have exactly the same lifetime, and also comparable masses. This was known as the θ–τ puzzle.[22]

At first, physicists, including Lee and Yang, attempted to solve this puzzle within the framework of conventional theories.[23] We shall discuss these attempts in a later section when we consider the intellectual context in which the suggestion of parity nonconservation was made. It is sufficient at this point to note that those attempts were unsuccessful. Lee and Yang saw clearly that a possible solution to the problem would be nonconservation of parity in the weak interactions, particularly in the decay of K mesons. Thus, although the K meson was produced in a state of definite parity, it could decay into either a two-pion state or a three-pion state, of opposite parity, and the τ and θ are merely different decay modes of the same particle. As Lee stated,

It finally became clear to me, in early 1956, that the solution to the θ–τ puzzle must lie in something much deeper: perhaps, parity is not conserved, and θ and τ are indeed the same particle. The immediate reaction to this proposal was "so what?". Unless it could be shown that parity conservation also extended to other processes, one would never know for sure whether in θ–τ decays parity conservation was, or was not, violated. Thus it was necessary to investigate how possible parity nonconservation effects could be observed in other weak reactions.[24]

In early 1956, Lee and Yang began to examine the existing evidence for the conservation of parity in previous experiments. They found, to their surprise, that all the earlier experiments in weak-interaction physics provided no evidence in support of conservation of parity.[25] They also found that although experiments on the strong and electromagnetic interactions established conservation of parity to a high degree of accuracy in those interactions, they were not sufficiently accurate to observe any effects of nonconservation of parity in the weak interactions. Lee reported their search as follows:

I then borrowed from Wu the authoritative book on β-decay edited by K. Siegbahn,[26] and proceeded with Yang to calculate systematically all parity violating effects. . . . After we went through the entire Siegbahn book, rederived all these old formulas with this new interaction, it became obvious to us that not only was there, at that time, not a single evidence for parity conservation in β-decay but that we must have been very stupid! . . . Once we stopped calculating and started to think, in a rather short time, it dawned on us that the reason for this lack of evidence was the simple fact that nobody had made any attempt to observe a physical pseudo-scalar[27] from an otherwise seemingly right–left symmetrical arrangement.[28]

I might add, parenthetically, that Lee and Yang did, in fact, fail to discover the two earlier experiments by Cox and associates[29] and by Chase[30] that might have given some evidence for nonconservation of parity. This is not surprising, because they were looking at experiments on β decay and weak interactions, whereas the earlier experiments were on the subject of electron scattering, which made use of the electrons from β decay only as the source of their beam.[31] Although these papers were known in the field of electron scattering, they virtually disappeared from mention in the physics literature in the late 1930s. We shall discuss these experiments and the reasons for their neglect in the next chapter.

Having found that there was no evidence in favor of conservation of parity in the weak interactions, and that nonconservation of parity would solve one of the most vexing problems in physics, the θ–τ puzzle, they were free to make their revolutionary suggestion. Yang indicated their state of mind as follows:

The fact that parity conservation in the weak interactions was believed for so long without experimental support was very startling. But what was more startling was the prospect that a space–time symmetry law which the physicists have learned so well may be violated. This prospect did not appeal to us. Rather we were, so to speak, driven to it through frustration with the various other efforts at understanding the θ–τ puzzle that had been made.[32]

Whether driven by frustration or not, Lee and Yang put forth their hypothesis in a paper, "Question of Parity Conservation in Weak Interactions,"[33] that is now justly regarded as a classic. The abstract is a masterpiece of understatement. "The question of parity conservation in β-decays and in hyperon and meson decays is examined. Possible experiments are suggested which might test parity conservation in these interactions."[34] This hardly seems to describe a paper that would undermine a long and strongly held principle of physics and that would ultimately lead to great theoretical advances in our understanding of the weak interactions. The introductory section gave both the problem and its solution:

Recent experimental data indicate closely identical masses and lifetimes of the $\theta^+(\equiv K^+_{\pi 2})$ and the $\tau^+(\equiv K^+_{\pi 3})$ mesons. On the other hand, analyses of the decay products of the τ^+ strongly suggest on the grounds of angular momentum and parity conservation that the τ^+ and θ^+ are not the same particle. This poses a rather puzzling situation that has been extensively discussed.

One way out of the difficulty is to assume that parity is not strictly conserved, so that θ^+ and τ^+ are two different decay modes of the same particle, which necessarily has a mass value and a single lifetime. We wish to analyze this possibility in the present paper against the background of the existing experimental evidence of parity conservation. It will become clear that existing experiments do indicate parity conservation in strong and electromagnetic interactions to a high degree of accuracy, but that for the weak interactions (i.e., decay interactions for the meson and hyperons, and various Fermi interactions) parity conservation is so far only an extrapolated hypothesis unsupported by experimental evidence. (One might even say that the present $\theta-\tau$ puzzle may be taken as an indication that parity conservation is violated in the weak interactions. This argument is, however, not to be taken seriously, because of the paucity of our present knowledge concerning the nature of the strange particles. It supplies rather an incentive for an examination of the question of parity conservation.)[35]

One might suppose that at some time in the future, a bright undergraduate, looking back on this episode, might wonder what all the fuss was about. Such a student might conclude that the decays of the θ^+ and τ^+ clearly demonstrated nonconservation of parity in the weak interactions. Nevertheless, Lee and Yang are clearly correct. As we shall discuss in a later section, the situation in the field of the so-called strange particles was so confused and these new developments so unexpected, including not only the $\theta-\tau$ puzzle but also other aspects of the field, that no one could say that anyone understood very much. In fact, as we shall see, among the proposed solutions to the $\theta-\tau$ puzzle, the hypothesis of parity violation was not only correct but also among the most plausible. Lee and Yang, however, did more than merely propose nonconservation of parity in the weak interactions. The

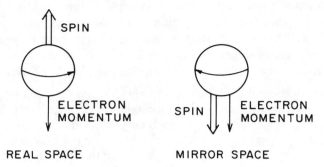

Figure 1.2   Spin and momentum in real space and mirror space.

possibility of such a violation had been mentioned earlier,[36] but only as a logical possibility; it had never been offered as the explanation of a physical problem. Lee and Yang also proposed several clear experimental tests of their hypothesis: "To decide unequivocally whether parity is conserved in weak interactions, one must perform an experiment to determine whether weak interactions differentiate the right from the left. Some such possible experiments will be discussed."[37] Although Lee and Yang proposed several experiments that were in fact later performed, we shall concentrate on only two, the β decay of oriented nuclei, and the sequential decay $\pi \rightarrow \mu \rightarrow e$, because these were the first experiments done, and they provided the crucial evidence for the physics community. We begin first with the β decay experiment:

A relatively simple possibility is to measure the angular distribution of the electrons coming from the β-decays of oriented nuclei. If θ is the angle between the orientation of the parent nucleus and the momentum of the electron, an asymmetry of distribution between θ and $180° - \theta$ constitutes an unequivocal proof that parity is not conserved in β decay.[38]

We can understand this as follows: Suppose we have an oriented or polarized nucleus with its spin pointing upward (Figure 1.2). Let us suppose that the electron from the β decay of this nucleus comes off in the direction opposite to the spin. If we perform a one-dimensional reflection, we see that the mirror image has the spin of the nucleus in the same direction as the electron momentum. Thus, the mirror image of the decay is different from the real decay, and parity is violated. The only experimental result that would show parity conservation would be the result in a group of oriented nuclei, with equal numbers of electrons coming off in the two directions. In that case, the mirror image of the group of decays would be identical with the real experiment. In reality, the situation is somewhat more complicated, because the electrons are emitted not only along the spin direction, but also with an asymmetrical angular distribution. As Lee and Yang pointed out,

To be more specific, let us consider the allowed β transition of any oriented nucleus, say Co^{60}. The angular distribution of the β radiation is of the form

$$I(\theta) = (\text{constant})(1 + \alpha \cos\theta) \sin\theta d\theta$$

where α is proportional to the interference term. . . . If $\alpha \neq 0$, one would then have a positive proof of parity nonconservation in β decay. . . . Thus, this experiment may prove to be quite feasible.[39]

This is precisely the experiment that was performed by Wu, Ambler, Hayward, Hoppes, and Hudson.[40] Their experiment consisted of a layer of polarized Co^{60} nuclei and a single electron counter that was located in a direction either parallel or antiparallel to the orientation of the nuclei. The direction of the polarization of the Co^{60} nuclei could be changed, and any difference in the counting rate in the fixed electron counter could be observed. Their results are shown in Figure 1.3 and clearly indicate the presence of an asymmetry, and thus parity violation. Their introduction mentions the recent proposal of Lee and Yang on the importance of this experiment to the question of parity conservation and states simply

If an asymmetry in the distribution between θ and $180 - \theta$ (where θ is the angle between the orientation of the parent nuclei and the momentum of the electrons) is observed, it provides unequivocal proof that parity is not conserved in β decay. This asymmetry has been observed in the case of oriented Co^{60}.[41]

This was a very difficult experiment, because it involved extremely low temperatures to preserve the orientation of the nuclei, the placing of a β counter inside the cryostat, and the location of the radioactive nuclei on a thin surface layer. The effect of temperature is shown clearly in Figure 1.3. As the sample warms up, the polarization of the nuclei is destroyed, and the asymmetry disappears.

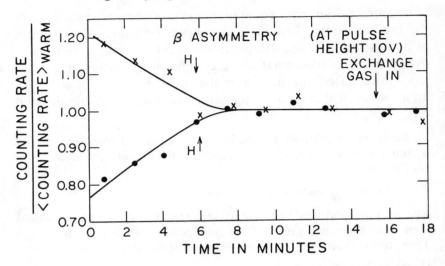

Figure 1.3   Relative counting rates for β particles from the decay of oriented Co⁶⁰ nuclei for different nuclear orientations (field directions). From Wu et al.[3]

The second experiment proposed by Lee and Yang was the decay sequence $\pi \to \mu \to e$:

In the decay processes

$$\pi \to \mu + \nu \tag{5}$$
$$\mu \to e + \nu + \nu \tag{6}$$

starting from a π meson at rest, one could study the distribution of the angle θ between the μ-meson momentum and the electron momentum, the latter being in the center of mass of the μ meson. If parity is conserved in neither (5) nor (6), the distribution will not in general be identical for θ and $\pi - \theta$. To understand this, consider first the orientation of muon spin. If (5) violates parity conservation, the muon would be in general polarized in its direction of motion. In the subsequent decay (6) the angular distribution problem with respect to θ is therefore closely similar to the angular distribution problem of β rays from oriented nuclei, which we have discussed before.[42]

This experiment was performed with two different techniques by Garwin, Lederman, and Weinrich[43] and by Friedman and Telegdi.[44] The experiment of Garwin and associates consisted of stopping μ mesons from π-meson decay in a block of carbon and detecting the electrons emitted in the subsequent decay of the μ

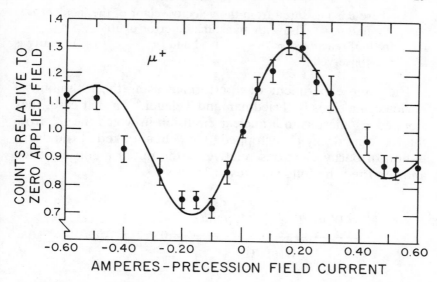

Figure 1.4 Relative counting rate as a function of precession field current. From Garwin et al.[4]

meson. Rather than move the electron counter to try to observe the asymmetry in the decay angular distribution, they found it easier to rotate the spin, or magnetic moment of the muons, and observe the electrons with a fixed counter:

Assume now that the processes of slowing down, stopping, and the microsecond of waiting do not depolarize the muons. In this case, the electrons emitted from the target may have an angular asymmetry about the polarization direction, e.g. for spin 1/2 of the form $1 + \alpha \cos\theta$. In the absence of any vertical magnetic field the counter system will sample this distribution at $\theta = 100°$. We now apply a small vertical field in a magnetically shielded enclosure about the target, which causes the muons to precess at a rate of $(\mu/s\hbar)H$ rad/sec. The probability distribution angle is carried around with the μ spin. In this manner we can, with a fixed counter system, sample the entire distribution by plotting counts as a function of magnetizing current for a given time delay.[45]

Their results are shown in Figure 1.4, clearly demonstrating the presence of an asymmetry. Their conclusions are stated as follows:

   I.   A large asymmetry is found for the electrons in $[\mu^+ \rightarrow e^+ + 2\nu]$, establishing that our μ^+ beam is strongly polarized.

  II.   The angular distribution of the electrons is given by $1 + \alpha \cos\theta$, where θ is measured from the velocity vector of the incident μ's. We find $\alpha = -1/3$ with an estimated error of 10%.

III. In [the] reactions $[\pi^+ \rightarrow \mu^+ + \nu]$ and $[\mu^+ \rightarrow e^+ + 2\nu]$ parity is not conserved.[46]

The same experiment was performed using the technique of nuclear emulsions by Friedman and Telegdi.[47] In that case, they stopped π^+ mesons in a nuclear emulsion in which the μ^+ from the resulting decay also stopped. They then looked at the distribution of decay electrons relative to the initial muon direction and reported the following result.[48]

$$\frac{[\int_{90°}^{180°} W(\theta) \, d\Omega - \int_{0°}^{90°} W(\theta) \, d\Omega]}{\int_{0°}^{180°} W(\theta) \, d\Omega} = 0.062 \pm 0.027$$

A note added in proof, based on further work, changed this result to 0.091 ± 0.022.

They, too, concluded that parity was not conserved in these decay processes.

Those experiments not only showed nonconservation of parity in weak interactions but also demonstrated violation of yet another law, that of charge-conjugation invariance, or particle–antiparticle invariance. In a paper that had not yet been published, but whose ideas were widely known, Lee, Oehme, and Yang[49] had shown that violation of parity conservation necessarily implied violation of charge-conjugation invariance.[50] All three of the experiments mentioned earlier stated the violation of that principle.

Soon there were many other experiments[51] that showed violation of parity conservation in the weak interactions. Of particular interest are the experiments that showed longitudinal polarization of electrons resulting from nuclear β decay. These were, in fact, quite similar to, and in one case virtually identical with, the experiments performed in 1928 and 1930.[52] This effect had also been predicted by Lee and Yang[53] in a later paper, and it was first demonstrated by Frauenfelder and associates.[54] Briefly, that experiment consisted in transforming the longitudinal polarization of the electrons to a transverse polarization and then detecting the transverse polarization by left–right asymmetry in the scattering of the electrons:

The observation of the expected longitudinal polarization is difficult. However, by means of an electrostatic deflector, the longitudinal polarization can be transformed into a transverse one. The transverse polarization can be measured by scattering the electrons with a thin foil of high-A material. (Mott scattering.) Because of the spin–orbit interaction, the elastically scattered electrons show a strong left–right asymmetry, especially at scattering angles between 90° and 150°. From this measureable asymmetry the initial longitudinal polarization can be calculated.[55]

A similar experiment was performed by Cavanagh and associates[56] in which the only major difference was that the transformation of the longitudinal polarization to transverse polarization was produced by crossed electric and magnetic fields.

From our point of view, the most important of the experiments on detection of electron polarization were the two done by Lipkin and associates.[57] Those experiments involved double scattering of electrons from two metal surfaces. The first scatter, at 90°, transformed the longitudinal polarization into transverse polarization. This was then detected on the basis of the left–right asymmetry in the second scattering. We shall postpone a discussion of the details of the physics involved until the next chapter, but we should note that these experiments were almost identical with the two prior experiments mentioned earlier.

1.2 CRUCIAL EXPERIMENTS?

We are now in a position to decide whether or not the experiments that demonstrated nonconservation of parity were crucial experiments in the sense that we discussed earlier. The first question we must answer is whether or not the breakdown of parity conservation is a sufficiently important discovery to merit the term "crucial." After all, there are experiments being done every day that may add to scientists' knowledge, or modify their views in a minor way, that certainly would not be called crucial.

It seems clear that although the breakdown of the theory of parity conservation did not cause as great a change in physicists' view of nature as did, for example, the change from Aristotelian to Newtonian dynamics or the change from Newtonian mechanics to relativity or quantum mechanics, it was sufficiently important and fundamental to be considered revolutionary. As we saw earlier, the indistinguishability of left and right was a feature of classical physics, and as we shall soon see, it was a widely and strongly held belief of the physics community from about 1927 (the publication date of Wigner's paper) to 1957, when the experimental results were announced. Perhaps the most telling evidence for the revolutionary character of this discovery is the fact that Lee and Yang were awarded the Nobel Prize for physics in 1957, probably the shortest period of time between a discovery and a Nobel award in history. O. B. Klein's presentation speech clearly indicated this:

But what has the question of left and right to do with elementary particle physics? Well, in the first place, only in a negative way, in that it was assumed almost tacitly, that elementary particle reactions are symmetric with respect to right and left. This assumption was to play an important part in the elaboration of Fermi's Theory. That this assumption was made very natural, not least in view of the mentioned theory of Dirac, according to which it looked as if the electrons, the best known elementary particles, possessed no feature which would permit a distinction between right and left. In fact, most of us were inclined to regard the symmetry of elementary particles with respect to right and left as a necessary consequence of the general principle of right–left symmetry of Nature. Thanks to Lee and Yang and the experimental discoveries inspired by them we now know that this was a mistake . . . nor could I do justice to the enthusiasm your new achievement has aroused among physicists. Through your consistent and unprejudiced thinking you have been able to break a most puzzling deadlock in the field of elementary particle physics where now experimental and theoretical work is pouring forth as the result of your brilliant achievement.[58]

We shall return to this when we discuss the evidence concerning the immediate acceptance of parity nonconservation by the physics community. The revolutionary nature of their discovery is also indicated in the hesitation of Lee and Yang to suggest their hypothesis without first conducting a detailed examination of the evidence in support of parity conservation, as discussed earlier.

Although it may not be considered as decisive evidence for the crucial nature of these experiments, or for the importance of the principle involved, it should be worth looking at the thoughts of these physicists before they performed their experiments. Wu reported her feelings as follows:

Following Professor Lee's visit, I began to think things through. This was a golden opportunity for a beta ray physicist to perform a *crucial test* [emphasis added], and how could I let it pass? Even if it turned out that the conservation of parity in beta decay was valid, the experimental result would, at least, set an upper limit on its violation and thus stop further speculation that parity is not violated.

That spring, my husband, Chia-Liu Yuan, and I had planned to attend an International Conference on High-Energy Physics in Geneva and then proceed to the Far East on a lecture tour. Both of us had left China in 1936, exactly twenty years earlier. Our passages were booked on the HMS Queen Elizabeth before I suddenly realized that I had to do the experiment immediately, before the rest of the physics community recognized the importance of this experiment and did it first. Although I felt that the chances of the parity conservation law being wrong were remote, I urgently wanted to make a clear-cut test. So I asked Chia-Liu to let me stay and go without me. Fortunately, he fully appreciated the importance of the time element and finally agreed to go alone.[59]

Professor Telegdi expressed a similar view:

Around August 1956 I ran across a preprint on Leona Marshall's desk: Lee and Yang's now classic article on parity violation and the $\theta-\tau$ puzzle. It instantly struck me as an extraordinarily exciting paper. It was obvious that the $\pi-\mu$ chain could readily be studied in emulsions; and I also felt that the parity violating effects could not be small if they were the explanation of the $\theta-\tau$ puzzle.

I gave a seminar on the Lee-Yang paper, vainly trying to communicate my excitement to others. I met with little enthusiasm from senior colleagues who tried to tell me that we would be wasting our time. I replied that I would be willing to risk wasting three months on such an exciting possibility.[60]

It is also important to examine the evidence that conservation of parity was a widely and strongly held belief in the physics community. There is evidence for this in the foregoing statements of Wu and Telegdi. Perhaps the earliest[61] evidence for this occurs in the work of Pauli (who was later to win a Nobel Prize himself) in his 1933 rejection of Weyl's two-component theory of spin-½ particles.[62] Pauli pointed out that Weyl's theory gave a mass of zero for the particles and also was not invariant under reflection: "However, as the derivation shows, these wave equations are not invariant under reflections (interchanging left and right) and thus are not applicable to physical reality."[63]

It is interesting to note that the theory that Pauli rejected because it violated parity conservation is quite similar to the theory later proposed by Lee and Yang,[64] Landau,[65] and Salam[66] that incorporated parity nonconservation into the theory of weak interactions in a very natural way.

Even as late as 17 January 1957, Pauli still had not given up his belief. In a letter to Victor Weisskopf, he wrote that "I do not believe that the Lord is a weak left-hander, and I am ready to bet a very large sum that the experiments will give symmetric results."[67] In a letter to C. S. Wu dated 19 January 1957, after hearing word of the results of her experiment, he noted that "I did not believe in it [parity nonconservation] when I read the paper of Lee and Yang."[68]

Frauenfelder and Henley were previously quoted as stating that parity conservation quickly became a sacred cow. They added later that the suggestion was "rejected by most physicists." They also reported an interesting anecdote: "R. P. Feynman [yet another Nobel Prize-winning physicist], for instance, bet N. F. Ramsey $50 to $1 that parity is conserved. Feynman paid."[69] Professor Lee also reported that Felix Bloch, another Nobel Prize winner, offered to bet other members of the Stanford Physics Department his hat that parity was conserved. He later remarked to Lee that it was fortunate he didn't own a hat.[70]

There is further evidence for the idea that parity conservation was a widely held belief in the textbooks written before 1957. Although I have not examined every such text, I have looked at a substantial number of them, and it seems fair to assume that every textbook on elementary-particle physics, nuclear physics, and quantum mechanics written prior to 1957 contains a statement of parity conservation. We can take as typical the statement in Halliday's *Introductory Nuclear Physics*, a textbook that was in wide use during the 1950s. In particular, it was the text used in Wu's course on nuclear physics given at Columbia.

Parity is an interesting property because of the law of conservation of parity. This states that the parity of an isolated system cannot change, no matter what transformations or recombinations take place within it. Suppose, for example, that a nucleus in an excited state is described by a wave function of even parity. If it emits a gamma ray and goes over to a lower energy state the system "recoiling nucleus + gamma ray" must continue to have even parity. This imposes some restrictions on the gamma-ray emission process; the conservation of parity, in other words, has led to some selection rules. Note that the parity of the nucleus above may or may not change. Similar parity restrictions are placed on radioactive-disintegration processes and on nuclear reactions.[71]

The next question to consider is whether or not this discovery was immediately accepted by the physics community. We saw earlier that Lee and Yang were awarded the Nobel Prize in 1957, which is certainly strong evidence in favor of immediate acceptance. Further evidence is provided by the fact that on 15 January 1957, the Columbia Physics Department held an unprecedented press conference to announce the discovery. At that conference, Professor I. I. Rabi, another Nobel Prize winner, said that "In a certain sense a rather complete theoretical structure has been shattered at its base and we are not sure how the pieces will be put together."[72] When Pauli received preprints of the experimental results, even he accepted them immediately. He wrote the following in another letter to Weisskopf:

Now, after the first shock is over, I begin to collect myself. Yes, it was very dramatic. On Monday, the twenty-first, at 8 p.m. I was to give a lecture on the neutrino theory. At 5 p.m. I received the three experimental papers [those of Wu, Lederman, and Telegdi]. I am shocked not so much by the fact that the Lord prefers the left hand as by the fact that He still appears to be left–right symmetric when He expresses Himself strongly. In short, the actual problem now seems to be the question: Why are strong interactions right and left symmetric?[73]

The annual meeting of the American Physical Society held in late January 1957 provides even more evidence. Because the program for the meeting was prepared several months in advance, obviously there was no time to include the new discoveries of parity nonconservation in the original program. Thus, a post-deadline session was scheduled at which Yang, Telegdi, Lederman, and Wu spoke. Dr. Karl K. Darrow, the secretary of the society, reported the session as follows:

Even more astonishing was the smash hit scored for the first time in history by post-deadline session, during which – and on Saturday afternoon to boot – the largest hall normally at our disposal was occupied by so immense a crowd that some of its members did everything but hang from the chandeliers. This was because the blackboards and the grapevine had spread the news that some of the post-deadline papers would be devoted to the issue of nonconservation of parity, which had burst into public view exactly two weeks before.[74]

By the time of the spring meeting of the society, the field of parity nonconservation was so well established that a symposium was scheduled on the subject, with Lee, Wu, Garwin, Telegdi, and K. M. Crowe delivering the talks.

The discovery of parity nonconservation also had a dramatic effect on the actual work done by physicists. We discussed earlier[51] the subsequent experiments to confirm parity nonconservation. Sullivan, White, and Barboni[75] have studied this problem in great detail. They report a large increase in the number of theoretical articles published in the field of weak interactions in 1957 and 1958 immediately after the discovery. There was no similar effect among experimental articles, but that was to be expected, because there were more constraints on the number of experiments that could be done, such as funding difficulties and construction of apparatus. However, if we look at the numbers of experimental papers devoted to various topics in weak interactions, we find that in both 1958 and 1959 approximately 60 percent of these articles were devoted to parity and its closely related field, the $V-A$ theory of weak interactions.[76] White and his collaborators also show that the discovery caused large demographic changes in the field of weak interactions. The numbers of both experimental and theoretical authors entering the field, as shown by the dates of their first publications, show large rises in 1957 and 1958.[77] This was true both for those referred to as "transients," who published a total of fewer than three articles in the field, and for "professionals," whose publications numbered three or more articles.

The textbook evidence also strongly indicates rapid acceptance of the discovery by the physics community. Textbooks, of course, do not show instantaneous effects because it takes rather a long time to produce them, but they do indicate what is considered acceptable by the physics community. A typical postdiscovery comment was that of Brancazio:

The principle of mirror symmetry had been tacitly assumed by physicists and was never held in doubt. Then in 1956 two theoretical physicists,

Yang and Lee, noted that the principle had, in fact, never been tested experimentally. Within a few months three different experimental observations of violations of mirror symmetry were made.[78]

Although it may not be considered evidence for the importance of this discovery, it is interesting to mention the dispute over priority. Although all three articles appeared in Volume 105 of *Physical Review*, the papers of Wu and Garwin were published in the 15 February 1957 issue, and Telegdi's article appeared in the 1 March 1957 issue. The receipt date for the first two articles was 15 January 1957, and for the third, 17 January 1957. As Telegdi recounts the story,[79] his letter was originally rejected on the grounds that it was unclear and confusing. The letter was eventually accepted and published with the following footnote: "For technical reasons, this letter could not be published in the same issue as that of Garwin et al., *Phys. Rev.*, *105* (1957), 1415."[80] Telegdi felt so strongly that this delayed publication was an injustice that he resigned from the American Physical Society, the publisher of *Physical Review*, and has not since rejoined the society. All of the histories of this episode, however, including this one, grant equal credit to all three experiments.

The response of Samuel Goudsmit, the editor of *Physical Review*, to Telegdi's statements is quite interesting, because it casts light on the nature of the evidence that is acceptable in such an important experiment. After mentioning the technical difficulties of publication, Goudsmit stated that

The obstacle encountered by Dr. Telegdi was of a different nature. A careful study of his Letter shows that at the time he submitted it the work was unfinished. He obtained an asymmetry of 0.062 ± 0.027 with 1300 events, that is $(690 - 610)/1300$ with a standard deviation of 35 in the difference. The effect is only a little larger than two standard deviations. This should be compared with the overwhelming and compelling evidence presented in the two other Letters as seen especially clearly in their graphs. An effect of less than three standard deviations is quite insufficient in such an important and subtle experiment. A very similar more recent example is the Columbia–Stony Brook work of 1966[81] on the decay of the η-meson, where an asymmetry of 0.072 ± 0.028 in 1441 events was reported. This was widely heralded as proof of violation of charge conjugation invariance. Later experiments showed that there is no asymmetry in the η-decay.

Several weeks after submission, Dr. Telegdi added a note in proof to his letter, stating that a total of 2000 events had now given an asymmetry of 0.091 ± 0.022, that is (1091 − 909)/2000. This is quite an improvement. Note, however, that the 700 additional cases must have shown a surprisingly large asymmetry, namely (401 − 299)/700 or 0.146 ± 0.039, as compared to the original 0.062 ± 0.027. These large statistical errors (27%–43%) definitely prove the preliminary nature of the initial result submitted for publication.[82]

In view of Goudsmit's comments, it seems appropriate to examine, with hindsight, of course, the nature of the evidence presented in support of parity nonconservation. Telegdi gave a result of 0.091 ± 0.022, or a 4.14-standard deviation (S.D.) effect. The probability that his result is compatible with a symmetric result is 3.5 x 10^{-5}, or 3.5 parts in 100,000. This is certainly a convincing result. In Professor Wu's experiment, parity conservation would require that the counting rate with the magnetic field up equal the counting rate with the magnetic field down. If we look at Figure 1.3, we see that the difference between the two counting rates is approximately 0.40. If we assume that the size of the data points is a measure of the error in the points, because no other errors are given, we obtain a standard deviation in the difference of approximately 0.03. Thus, we have a 13-S.D. effect. The probability of this happening is so small that the standard compilations do not list it. We can take as a comparison the fact that a 10-S.D. effect has a probability of 1.5 parts in 10^{23}. This is clearly an overwhelming result. The results of the experiment of Garwin and associates are statistically even more convincing. As we see from Figure 1.4, the experimental results give a sinusoidally varying counting rate as a function of precession field current, whereas parity conservation would imply that the counting rate should be flat. Garwin reported that there was a 22-S.D. effect and wrote that "It was very pleasant to have such a prompt and unambiguous result."[83] When Lederman saw the results, he called Lee at about seven in the morning and announced that "Parity is dead."[84]

Thus, at least on statistical grounds, the evidence presented for nonconservation of parity in the three original experiments is nothing short of overwhelming. As Mark Corske, one of my former students, remarked, "Four standard deviations is strong evidence, thirteen is absolute truth, and twenty-two is the word of God." Although it is not likely that many physicists at the time calculated the probabilities, a mere glance at the data would certainly have been sufficient; it is clear that the decision by the physics community to accept parity nonconservation was a reasonable decision. The question how one comes to believe rationally in experimental results will be discussed in detail in Chapters 6 and 7, and, in particular, these experiments will be discussed.

I have presented, I believe, convincing evidence that the experiments of Wu, Ambler, Hayward, Hoppes and Hudson, and of Garwin, Lederman, and Weinrich, and of Friedman and Telegdi were crucial experiments. They decided unambiguously, within a short period of time, for the physics community, between two classes of theories: those that would conserve parity and those that do not.

1.3 THE REASONS WHY

Let us now examine the intellectual situation in the 1950s to try to see why Lee and Yang felt driven to make their suggestion of parity nonconservation and why it was taken seriously by the physics community. It was mentioned previously that the suggestion that parity might not be conserved had been made before Lee and Yang, but that it had been offered as a possibility rather than as a solution to a real problem. The first instance was in a paper dealing with the possibility of electric-dipole moments for elementary particles and nuclei by Purcell and Ramsey.[85] They noted that one of the major theoretical arguments against the existence of such electric-dipole moments rested on the conservation of parity, which had not yet been tested for elementary particles and nuclei:

It is generally assumed on the basis of some suggestive theoretical symmetry arguments that nuclei and elementary particles can have no electric dipole moments. It is the purpose of this note to point out that although these theoretical arguments are valid when applied to molecular and atomic moments whose electromagnetic origin is well understood, their extension to nuclei and elementary particles rests on assumptions not yet tested. . . .

The argument against electric dipoles, in another form, raises directly the question of parity. A nucleon with an electric dipole moment would show an asymmetry between left- and right-handed coordinate systems; in one system the dipole moment would be parallel to the angular momentum and in the other, anti-parallel. But there is no compelling reason for excluding this possibility. It would not be the only asymmetry of particles of ordinary experiences, which already exhibit conspicuous asymmetry in respect to electric charge.[86]

That paper was erroneously quoted by Lee and Yang. They stated that

By far the most accurate measurement of the electric dipole moment was made by Purcell, Ramsey, and Smith. They gave an upper limit for the electric dipole moment of the neutron of e x (5 x 10^{-20} cm). This value ... is also the most accurate verification of the conservation of parity in the strong and electromagnetic interactions.[87]

Their footnote 6 mentions the paper of Purcell and Ramsey as quoted in N. F. Ramsey, *Molecular Beams*. Although the original paper made no mention of any quantitative result for the electric-dipole moment of the neutron and only mentioned that the experiment was in progress, the book implies that the result is contained there.[88] It seems clear, therefore, that Lee and Yang never read the original paper, and Lee has confirmed this.[89]

There is a further suggestion of the possibility of parity non-conservation in a 1952 paper by Wick and associates.[90] They discussed the concept of intrinsic parity of elementary particles and pointed out that there were limitations to this concept:

The purpose of this paper is to point out the possible (and in certain cases necessary) existence of limitations to one of these general concepts, the concept of "intrinsic parity" of an elementary particle. Even though no radical modification of our thinking is thereby achieved, we believe that the injection of a certain amount of caution in this matter may be useful, as it may prevent one from calling "theorems" certain assumptions or from discarding as "impossible" forms of the theory, which under a more flexible scheme are perfectly consistent.[91]

They then proceeded to discuss some possible forms of the theory, with particular reference to both particle–antiparticle symmetry and inversion symmetry (parity conservation):

That C [particle–antiparticle symmetry] is an exact symmetry property is moreover far from proved. The disturbing possibility remains that C and I [inversion symmetry or parity conservation] are both only approximate and CI is the only exact symmetry law. ... This possibility, however, seems rather remote at the moment.[92]

As we have seen, the possibility was not quite as remote as Wick and associates believed, because it was only four years later, to the day, that Lee and Yang published their paper on parity nonconservation. By mentioning these earlier suggestions I do not in any way mean to imply that they influenced Lee and Yang or that they were precursors in any real sense. They show only that by 1950 there was at least a consideration of the possibility of such symmetry violations. It still was a crucial and major step not only to suggest that such violations would solve a physical problem, namely the $\theta-\tau$ puzzle, but also to discuss the experimental tests of such a suggestion, as Lee and Yang did.

In order to try to understand the context in which Lee and Yang made their bold suggestion, we shall now look at the situation during the 1950s with regard to the so-called strange particles. These particles were called strange because their behaviors were very peculiar, when compared with those of other elementary particles, and not very well understood. The study of these particles began in 1947 with the discovery of new unstable particles by Rochester and Butler.[93] It was found that these new particles were of two types: those that had a mass approximately half that of the proton, called K particles or K mesons, and those that were slightly heavier than the proton, called hyperons. The K mesons, in particular, presented a puzzle, because there seemed to be either (1) many particles with equal or nearly equal masses and lifetimes and different decay modes or (2) perhaps fewer particles, possibly only one, of which the observed K particles were merely different decay modes. These included the θ and τ mesons, which we have already discussed, as well as others.

In a review article published in 1956, A. M. Shapiro noted, that "During the years 1949 through 1954 at least eight different K mesons or K-particle decay modes and at least four different hyperons were discovered."[94] In another 1956 review of mesons and hyperons, M. M. Shapiro listed six separate particles or decay modes.[95] The question of how many different particles or decay modes seemed to be essentially a matter of taste, depending on whether or not one considered different charge states as separate modes.

Perhaps the most puzzling feature of the strange particles, aside from the $\theta-\tau$ puzzle, was the fact that although they were produced quite copiously in strong interactions, their lifetimes were quite long. Yang summed this up in a review talk given at the

Rochester conference on high energy nuclear physics in April 1956:

The starting point of these considerations was, as you remember, the puzzle that, while the strange particles are produced quite abundantly (say 5% of the pions) at BeV energies and up, their decays into pions and nucleons are rather slow (10^{-10} sec). Since the time scale of pion-nucleon interactions is of the order of 10^{-23} sec, it was very puzzling how to reconcile the abundance of these objects with their longevity (10^{13} units of time scale).[96]

The first suggestion on how to solve this problem was offered by Pais in 1952.[97] He suggested that in the strong interactions, the numbers of such particles produced were always even – the phenomenon of associated production. In the decay of these particles, however, because only one such particle was involved, the strong interactions would not operate, and the decays would proceed only through the weak interactions. In an ad hoc way, at least, this explained the phenomenon. This suggestion was elaborated in the theory of strangeness conservation, put forth by Gell-Mann and Nishijima during the period 1953–5.[98] They conjectured that the K mesons and hyperons possessed a quantity called strangeness (+1 for the K^0 meson and -1 for the Λ hyperon) that was conserved in the strong interactions, but not in the weak interactions. This is, of course, quite similar to Pais's suggestion and does explain both the phenomenon of copious associated production and that of slow decay of the strange particles. The theory of conservation of strangeness did, however, predict that certain other particle reactions would not occur, and in 1956 Yang stated that "This conservation of strangeness was proposed by Gell-Mann and by Nishijima in 1953, and during the past year it was given very strong experimental support."[99] After listing several experimental results that supported the hypothesis, Yang stated that "There are no known violations of the strangeness selection rule."[100]

The study of the puzzling phenomena of strange particles was clearly of major interest to physicists. During the period from 1950 to 1956 there was a dramatic increase in the number of experimental papers on weak interactions, and the percentages of such papers devoted to K-meson decay were approximately 50

percent in 1954, 65 percent in 1955, and 65 percent in 1956.[101]
During the same period, new developments in the design and
construction of high-energy accelerators and advances in the techniques of particle detection greatly facilitated such studies. As
Yang remarked in 1957,

Experimental work on these strange particles, and others found later,
rapidly expanded in the last few years. With the completion in 1953 of
the Cosmotron and in 1955 of the Bevatron, experiments entered into
a new era in which one can work with controlled beams of strange
particles and in which one is able to make relatively accurate measurements on their interactions and decay. The scope and intensity of the
experimental activities on this subject may be gauged by the fact that
at Brookhaven National Laboratory in the United States, 60% of the
Cosmotron time at present is devoted to the study of strange particles.[102]

This concentration on experimental study of the strange particles was, of course, accompanied by theoretical work. In particular, various attempts were made to solve the $\theta-\tau$ puzzle using
both conventional and unorthodox methods. The earliest of these
was carried out in 1955 by Lee and Orear, who suggested that
perhaps the heavier of the two particles might decay into the
lighter one, thereby giving a theory consistent with the experimental data:

From the analysis of Dalitz it is becoming necessary to assume that the
$K^{\pm}_{\pi3}(\tau^{\pm})$ and $K^{\pm}_{\pi2}(\theta^{\pm})$ are due to at least two different charged particles.
Recent experimental evidence encourages the hypothesis that the ob-
servable lifetimes are exactly the same (at least after travel times $>10^{-9}$
sec). As a solution to this possible dilemma we propose that there
exist two heavy mesons, the $\theta^{\pm}$ and the $\tau^{\pm}$, where $\theta^{\pm} \rightarrow \pi^+ + \pi^0$ and
$\tau^{\pm} \rightarrow \pi^{\pm} + 2\pi$. We propose further that the heavier of these two has
a lifetime of 10^{-8} sec with a significant branching ratio for gamma decay
to the lighter one. Finally, if we assign a lifetime of order 10^{-9} sec or
less to the lighter one, we achieve a situation where both particles have
different lifetimes, but in most experiments appear to have exactly the
same lifetime.[103]

At the 1956 Rochester conference, Alvarez reported on a
search for the γ rays that would arise from the proposed Lee and
Orear decay schemes, $\theta^{\pm} \rightarrow \tau^{\pm} + 2\gamma$ or $\tau^{\pm} \rightarrow \theta^{\pm} + \gamma$, and
concluded "Therefore, no Lee-Orear γ-rays were seen with pulses
of the order of 0.5 MeV or greater."[104]

Lee and Yang made another suggestion to explain the apparent equality of masses of the θ and τ.[105] They proposed that as a result of another symmetry, which they called parity conjugation, every strange particle occurs as a parity doublet. Thus, there would be not only the θ and τ, which have opposite parities, but also Λ_1^0 and Λ_2^0, and so forth. No evidence of such Λ^0 particles was seen. The subsequent observation of parity nonconservation eliminated this hypothesis.

The Rochester conference on high-energy nuclear physics in 1956 provided a large number of other theoretical proposals. Marshak[106] suggested that a last effort be made to explain θ and τ as a single particle using larger spin values. He reported that the lowest spin value compatible with the single-particle assumption was 2^+. Although this was compatible with the data existent at the time, Orear and associates[107] subsequently showed that this was extremely unlikely.

Following Yang's review talk on the subject of strange particles, there was an extensive discussion of various other alternatives. It is useful to examine the entire summary of that discussion, because it indicates clearly how physicists were thinking about the problem:

An extensive discussion followed. . . . Pursuing the open mind approach, Feynman brought up a question of (Martin) Block's: Could it be that the θ and τ are different parity states of the same particle which has no definite parity, i.e., that parity is not conserved: That is, does nature have a way of defining right-or left-handedness uniquely? Yang stated that he and Lee looked into this matter without arriving at any definite conclusions. Wigner (as discussed in some notes prepared by Michel and Wightman) has been aware of the possible existence of two states of opposite parity, degenerate with respect to each other because of space-time transformation properties. So perhaps a particle having both parities could exist. But how could it decay if one continues to believe that there is absolute invariance with respect to space-time transformations? Perhaps one could say that parity conservation, or else time inversional invariance could be violated. Perhaps the weak interactions could all come from this same source, a violation of space-time symmetries. The most attractive way out is the nonsensical idea that perhaps a particle is emitted which has no mass, charge, and energy-momentum, but only carries away some strange space-time transformation properties. Gell-Mann felt that one should also keep an open mind about possibilities like the suggestion by Marshak [see above] that the θ and

τ may, without requiring radical assumptions, turn out to be the same particle.

Michel suggested another way out of the difficulty. What is seemingly well known from experiment is the parity of the τ^+ and θ^0. If one assumes that the π^0 emitted in the process $\theta^+ \rightarrow \pi^+ + \pi^0$ is a "nearly real π^0" in the same sense that the pair emitted in the process $\pi^0 \rightarrow \gamma + e^+ + e^-$ is a "nearly real γ," there is nothing to prevent θ^+ from having spin-parity 0^-. This "nearly real π^0" would appear only virtually, perhaps due to some selection rules.

The chairman (J. R. Oppenheimer) felt that the moment had come to close our minds. . . . [108]

It was within this context, one of intense experimental and theoretical activity, of the failure of orthodox methods to explain the $\theta-\tau$ puzzle, and of some almost desperate attempts at explanation, that Lee and Yang offered their suggestion of parity nonconservation in the weak interactions. It comes as no surprise that at least some members of the physics community considered it seriously and acted on it, leading to the crucial experiments we discussed earlier.

1.4 DISCUSSION[109]

It was shown earlier that the results of those experiments were accepted as crucial by the entire physics community. It seems fair to say that as soon as physicists saw those results, they were agreed that parity was not conserved in the weak interactions.

There were several reasons for this. First, the suggestion of parity nonconservation solved a particularly vexing problem, namely, the $\theta-\tau$ puzzle, that had resisted all conventional attempts at solution. More important, however, was the fact that Lee and Yang made novel predictions that were then confirmed by overwhelming statistical evidence, as we have seen. We might also note that other predictions by Lee and Yang were subsequently confirmed, although by the time they were confirmed, the issues involved had been decided. It is possible, however, to ask if the experimental results in the parity question might have been in error – that despite the overwhelming statistical evidence there might have been systematic errors in the apparatus that gave rise to results that seemed to show that parity was not conserved, when in fact it was. This was extremely unlikely in view of the magnitude of the effects seen and also because of the fact that there were three independent results from two separate experiments, using three vastly different techniques. More detailed discussion of how we characterize different experiments and why they give us more confidence in our results will be presented in Chapters 4 and 6.

We may still ask if it might have been possible to modify background knowledge or use some auxiliary hypothesis to preserve parity conservation, in line with suggestions made on general grounds by Duhem and Quine.[110] It is interesting to note, however, that a survey of the literature of the time indicates that no attempts were made to do this[111] and that theoretical work was concentrated on incorporating parity nonconservation into the theory of weak interactions in a natural way, culminating in the successful *V–A* theory. This is not, however, an answer to the problem raised by Duhem and Quine. We might ask if attempts could have and should have been made to preserve the concept of parity conservation in the weak interactions.

In this context it is important to consider the kind of question being asked. Lee and Yang had stated that observation of the expected effects would constitute "unequivocal proof that parity is not conserved," and they were correct. We are dealing here not with two competing theories that make different quantitative predictions but with two different classes of theories: those that satisfy a particular symmetry (in this case, parity conservation) and those that do not. These classes of theories are, in fact, both mutually exclusive and exhaustive. The answer to the question whether or not a particular symmetry is satisfied is either "yes" or "no." This is not, of course, to deny that quantitative considerations can enter into such a decision, because the magnitude of the effect may be such that the experimental results will be inconclusive. There seems, nevertheless, to be a distinction between the question whether or not there is a specific condition that must apply to all theories concerning certain phenomena and the question which theory gives numerical results in agreement with experiment. The successful *V–A* theory of weak interactions is only one of many possible parity-violating theories, but there are only two classes of such theories: those that conserve parity and those that do not. Once one accepts the evidence in favor of the observed asymmetry, then it seems that one must either attribute it directly to parity nonconservation in the weak interactions or else say that parity is conserved in the weak interactions and modify the background beliefs so that the universe itself is asymmetric. There does not seem to be any alternative that preserves both parity conservation in the weak interactions and the symmetry of the universe, given the correctness of the experimental results. If a modification of the background beliefs had been made, then the experiments would have shown conclusively that parity is violated in nature – a conclusion that is of similar interest and importance as the conclusion that parity is violated in the weak interactions. Duhem and Quine did allow the whole of science to be refuted, although they argued that the refutation could not be localized to any particular theory or hypothesis. I have argued here that it is the very general nature of symmetry principles that allows them, and parity conservation in particular, to be refuted even though they are not the whole of science.

It is also quite important that the physics community had previously been able to classify the interactions in nature into four classes (strong, electromagnetic, weak, and gravitational) and that Lee and Yang's suggestion applied only to the weak interactions. They had already noted that there was convincing evidence in favor of parity conservation in the strong and electromagnetic interactions, but absolutely no evidence either way for the weak interactions. That certainly helped to define the issue, because the community was not faced with an effect seen in some reactions, involving a single class of interaction, but not seen in other reactions; rather, it was faced with an effect that was to be seen in all weak interactions.[112]

We should also note that the suggestion of parity nonconservation in the weak interactions cast absolutely no doubt on the operation of the apparatus involved in the experiment. The apparatus used depended only on the strong and electromagnetic interactions, in which parity conservation was already well corroborated. That avoided the danger of a vicious circularity analogous to that encountered in trying to measure the temperature dependence of a volume of material using a mercury thermometer as a measure of temperature when the operation of the thermometer itself depends on the hypothesis one wishes to test. Although the apparatus was laden with a theory, it was not the theory under test.

There is yet another possible factor in the decision that appears to have been generally discounted by physicists. Even though they discounted this factor, I believe that it must have had at least an indirect effect on their thinking. It was the example of a conservation law that held for the strong and electromagnetic interactions but was violated in the weak interactions, namely, the conservation of strangeness, discussed previously. Thus, there was at least a precedent for a conservation law that was violated in one class of interactions. We note here the fact that the case of strangeness conservation is slightly different from that of parity nonconservation. It was known at the time that if strangeness was to be a quantity of physical interest, it would not be conserved in the weak interactions. The question to be answered was whether or not it was conserved in the strong and electromagnetic interactions.

Nonconservation of parity also implied violation of charge-conjugation or particle-antiparticle symmetry, a fact that was noted in all three experimental papers. It can be suggested that acceptance of parity nonconservation was made easier by associating it with a change in structure (i.e., particles and antiparticles).[113] It was believed, at the time, that the combined operation (parity and charge conjugation) was a good symmetry. The discovery of the violation of this CP symmetry will be discussed in Chapter 3.

2

The nondiscovery of parity
nonconservation

In the previous chapter we mentioned two earlier experiments, performed by Cox, McIlwraith, and Kurrelmeyer, "Apparent Evidence of Polarization in a Beam of β-Rays," and by Chase, "The Scattering of Fast Electrons by Metals. II. Polarization by Double Scattering at Right Angles,"[1] that, at least in retrospect, provided evidence for nonconservation of parity in the weak interactions.[2] The anomalous nature of those experimental results seems to have been fairly well known (we shall discuss this later), although the exact nature of the anomaly was not clear. However, one thing is certain: The relation of those results to the principle of conservation of parity was not recognized or understood by any contemporary physicists, including the authors themselves.[3] Before discussing the reasons leading to this conclusion, it is worth examining in some detail what those results were, how they were obtained, and whether or not they really provided evidence for nonconservation of parity.

2.1 DID THE EXPERIMENTS SHOW NONCONSERVATION OF PARITY?

The intellectual context for those experiments involved the desire to demonstrate the vector nature of electron waves. This had started with the suggestion of de Broglie[4] in 1923 that just as light exhibited both particle and wave characteristics, so should those things that we normally consider particles exhibit wave characteristics. This had been brilliantly confirmed in 1927 in an experiment on the diffraction of electrons by crystals performed by Davisson and Germer.[5] This idea of electron waves was then combined with the concept of electron spin, which had been used by Uhlenbeck and Goudsmit[6] to explain certain features of atomic spectra. The mathematics of a vector electron were then worked out by Darwin.[7] Cox and his collaborators thought that an experiment in which electrons were double scattered would provide evidence for this, in analogy with experiments on light and X rays in which the first scattering polarized the light and the second scattering acted as an analyzer:

The already classic experiment of Davisson and Germer in which the diffraction of electrons by a crystal shows the immediate experimental reality of the phase waves of de Broglie and Schrödinger suggested that it might be of interest to carry out with a beam of electrons experiments analogous to optical experiments in polarization. It was anticipated that the electron spin, postulated by Compton to explain the systematic curvature of the fog-tracks of β-rays, and recently so happily introduced in the theory of atomic spectra by Uhlenbeck and Goudsmit might appear in such an experiment as the analogue of a transverse vector in the optical experiments. This idea has lately been developed by Darwin.[8]

Although the general nature of the effect to be observed in this experiment was known from the optical analogies, the detailed calculation of the effect was not carried out until 1929, by Mott.[9] Mott recognized that use of the double scattering of electrons was one of the few methods available to observe the spin of the free electron, because direct measurements would fail for technical reasons. It is worth examining Mott's discussion, even though it was not available to Cox and his collaborators (it was known to Chase), because it provides a very useful discussion of the physics involved. Lee Grodzins, the first person to recognize the significance of these early experiments, summarized Mott's discussion of the scattering of an initially unpolarized electron beam as follows:

In 1929, Mott showed that the spin orbit coupling in Coulomb scattering could be used as a polarizer or analyzer of electron spin. The spin orbit coupling is largest when the energy of the electrons is relativistic, the Z of the scatter is high, and large angle scattering occurs: most calculations consider the single scattering through 90° by nuclei of $Z \cong 80$. Mott double scattering is illustrated in [Figure 2.1], where an unpolarized beam of relativistic electrons is first polarized and then analyzed by being scattered singly twice by very thin high Z scatterers; $\bar{n}_1$, is a unit vector in the direction from the source to scatter 1, $\bar{n}_2$ is a unit vector in the direction of scatter 2, and $\bar{n}_3$ is a unit vector in the direction of the detector. The spin orbit term in the cross section is of the form $\bar{\sigma} \cdot (\bar{n}_1 \times \bar{n}_2)$ so that only a spin perpendicular to the plane of scattering

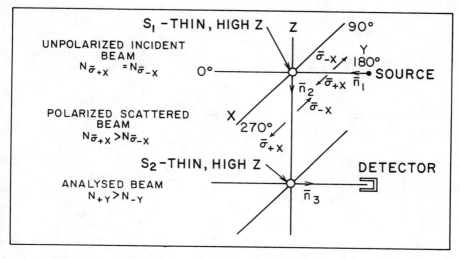

Figure 2.1. Mott double scattering for an unpolarized beam. From Grodzins.[2]

is affected. The single scattered beam at $\bar{n}_2$ will be partially polarized in the $+X$ direction. In the second scattering the spin orbit part of the cross section $\bar{\sigma}_{+x} \cdot (\bar{n}_2 \times \bar{n}_3)$, will add to the Coulomb term, while $\bar{\sigma}_{-x} \cdot (\bar{n}_2 \times \bar{n}_3)$ will subtract, so that a greater counting rate will be observed when the source is at the 180° position compared to that observed at the 0° position. No difference in counting rate should be observed when the 90° and 270° positions are compared.[10]

The same argument applies if the source is held fixed and the detector is rotated. We shall now consider the effects to be observed if the initial electron beam is longitudinally polarized (i.e., the spin is preferentially either along the direction of the electrons' motion or opposite to that motion). Although this situation was not treated by Mott, it is of crucial importance in understanding the experimental results of Cox and his collaborators and Chase. The very existence of a longitudinal polarization for electrons from β decay is evidence for nonconservation of parity. This is made clear by reference to Figure 1.2. In this case we regard the spin as the spin of the electron itself, rather than that of the nucleus. Let us assume that the electron spin is opposite to its momentum. A one-dimensional mirror reflection will change the spin direction, but not the momentum, and the mirror image will

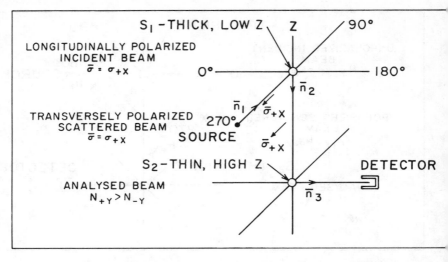

Figure 2.2. Double scattering for a longitudinally polarized incident beam. From Grodzins.[2]

have the spin in the same direction as the momentum, clearly showing a difference. This longitudinal polarization can also be detected in a double-scattering experiment. Grodzins again provides a summary of the physics:

The double scattering of an initially longitudinally polarized beam also results in strong asymmetries. However, the term $\bar{\sigma}_1 \cdot (\bar{n}_1 \times \bar{n}_2)$ is zero so that the asymmetry at 0° and 180° is small and the main effect is observed between the 90° and 270° positions. A double scattering experiment on beta rays in which both the first and second scatterers are so thin that only single scattering takes place is very difficult. No such experiment has been reported. Experimenters have, however, measured beta helicities [longitudinal polarizations] by first transforming the polarization from longitudinal to transverse and then utilizing Mott scattering. The transformation may be accomplished by an electric field or by multiple scattering, since small angle scattering is spin independent. In the latter case, if the first scatterer is a thick, preferably low Z, material, the spin direction will be unchanged as the momentum turns, and the beam at n_2 will be transversely polarized in the n_1, n_2 plane, along the X direction in [Figure 2.2]. When the beam is scattered through 90° by a thin high Z scatterer, the large spin orbit term will lead to different intensities along the Y axis, or for a fixed detector, a difference in intensities for the source positions at 90° and 270°. . . . [11]

With this background in physics we are now in a position to discuss these earlier experiments. Cox and his colleagues described their experiment as follows:

In our experiment β-particles, twice scattered at right angles, enter a Geiger counter. The relative numbers entering are noted as the angle between the initial and final segments of the path is varied. For reasons to be mentioned later, the angles at which most of the observations have been made are indicated [Figure 2.2] as 270° and 90°. The difference between the configurations of the three segments of path at these two angles is the same as the *difference between right- and left-handed rectangular axes* [emphasis added].[12]

Their targets consisted of gold plugs, and a milligram of radium was used as the source of β particles. The scattered β particles were then detected by platinum-point Geiger counters, which were quite unreliable and also were sensitive to background electrons:

Discharges are produced not only by β-particles, but also by photoelectrons ejected from the apparatus by the γ-rays of the radium. The high penetration of these rays makes it impossible to shield against them without interposing so much material that the path of the β-particles would be too much lengthened. . . . Their numbers however, could not be neglected, but there is no reason to expect that it would vary between the two settings at which most of the counts were made.

Although the platinum points described were found the best of several types and materials that were tried, they are far from satisfactory. They usually gave inconsistent results after an hour or two of use and have to be replaced. Moreover, the counts obtained with two different points do not agree. For this reason and on account of the uncertainty of the effect of the γ-rays, it seemed inadvisable to attempt counts all around the circle. Attention was given instead to taking counts to test an early observation that fewer β-particles were recorded with radium at 90° than at 270°.[13]

Their experimental results are given in Table 2.1. The weighted average of their results gives the ratio of the count at 90° to the count at 270° as 0.91 ± 0.01. They noted, however, that their results varied considerably among the different runs:

It will be noted that of these results a large part indicate a marked asymmetry in the sense already mentioned. The rest show no asymmetry beyond the order of the probable error. The wide divergence among

Table 2.1. *Experimental results of Cox and associates*[a]

Count at 90° / Count at 270°	0.76	0.90	0.94	0.87	0.98	1.03	1.03	0.91
Probable error	0.01	0.07	0.01	0.02	0.01	0.03	0.02	0.02
Count at 90° / Count at 270°	0.95	0.99	1.01	1.06	1.05	0.55	0.91	
Probable error	0.05	0.03	0.04	0.05	0.02	0.05	0.03	

[a]Weighted average 0.91 ± 0.01. (This average was not calculated in the original paper.)

the results calls for some explanation, and a suggestion to this end will be offered later. Meanwhile, a few remarks may be made on the qualitative evidence of asymmetry. Since the apparatus is symmetrical in design as between the two settings at 90° and 270°, the source of the asymmetry must be looked for in an accidental asymmetry in construction or in some asymmetry in the electron itself.[14]

They then examined the possible sources of systematic error in their experiment, such as lack of proper centering of the radioactive source and Geiger counter, some asymmetry in target orientation, the possibility of some residual magnetic field in the apparatus, and the possibility that the electron was polarized in passing through some material in the apparatus. They rejected all of these as unlikely. They then concluded:

It should be remarked of several of these suggested explanations of the observations that their acceptance would offer greater difficulties in accounting for the discrepancies among the different results than would the acceptance of the hypothesis that we have here a true polarization due to the double scattering of asymmetrical electrons. This latter hypothesis seems the most tenable at the present time.[15]

These authors offered no theoretical explanation of their results, but they did suggest that the discrepancies in their results might be attributable to a velocity dependence in their Geiger-counter efficiencies:

The discrepancies observed we ascribe to a selective action in the platinum points, whereby some points register only the slower β-particles. Observations in apparent agreement with this assumption have recently been made by Riehl. It is necessary to suppose further that the polarization is also selective, the effect being manifest only in the faster β-particles.[16] Perhaps the simplest assumption here is that only β-particles which are scattered without loss of energy show polarization.[17]

Before examining the validity and meaning of these results, it is worth examining the continuation of the experiment carried out by Carl Chase, then a graduate student working under the supervision of Cox. Chase's first experiment[18] started with an apparatus that was quite similar to that of Cox and associates, the major difference being that the targets were made of lead rather than gold. He obtained in the early data "an asymmetry of the same nature as had been observed by the earlier three, the effect being in the same direction and on the average of the same magnitude."[19] There was, however, a large background due to γ rays that had also been present in Cox's experiment, and the apparatus was redesigned to eliminate this background. His new results were as follows:

Azimuthal angle	Relative count
0°	1.000
90°	0.977
180°	0.958
270°	0.969

There was an error of less than 1 percent. He stated his conclusion as follows:

It will be seen that there is no indication of any effect of polarization either of the kind suspected by Cox, McIlwraith, and Kurrelmeyer, or of a kind occurring in Barkla's X-ray experiment. . . .
One must conclude then that a beam of electrons is not polarized by scattering, even when the electrons have the high speeds of the beta rays from radium E. It is hard to see how an asymmetry in the apparatus of Cox, McIlwraith, and Kurrelmeyer, or in the earliest apparatus of the present writer, could have produced the observed effect in such a large fraction of the number of runs, because of the care used in making the apparatus symmetrical, and in inverting the radium, etc. In the writer's case, this effect must have been entirely due to gamma-rays, in which case the asymmetry must have been due to some asymmetry in the apparatus. . . . But in the final work herein discussed, electrons alone are counted and there is no effect. It may be remarked that the apparatus of Cox et al. and that first used by the present writer which gave similar results, were both made on the same lathe.[20]

By this time, Mott's first paper on electron scattering, which predicted an asymmetry between the 0° data and the 180° data, had
already been published. Chase, who had observed such an asymmetry, attributed it to a difference in path rather than to a polarization effect.

Chase, however, continued his work, including a survey of the
literature, and he noted that

> All experiments with slow electrons have given negative results, while
> some of the experiments with fast electrons have shown evidence of
> polarization effects. As far as any theory is available, the prediction is
> that the spin vector of the electron may possibly show itself as the analog
> of a transverse vector in the electron waves; but only when the electrons
> have high velocities.[21]

He then proceeded to examine the velocity sensitivity of the
Geiger counters, an effect that had been noted earlier by Cox.
He concluded that the sensivity of the counters did depend quite
strongly on the counter windows and the voltage applied and that
this cast some doubt on his own previous work:

> From the results contained in this paper, it appears that this later work
> [his earlier paper] did not include a count of the faster electrons. Thinner
> windows allowed more slow electrons to get through. The increased
> number of electrons in turn allowed the operation of the counter at
> lower voltages, making for more consistent results, but counting only
> the slower electrons. . . . The foregoing shows the need for further work
> on these experiments which we have undertaken.[22]

In view of the difficulties involved with the use of Geiger
counters, Chase continued his work, but this time with an electroscope used as a detector.[23] His results are given in Table 2.2.
He obtained a result for the ratio (count at 90°)/(count at 270°)
= 0.973 ± 0.004. He concluded:

> The following can be said of the present experiments: the asymmetry
> between the counts at 90° and 270° is always observed, which was in no
> sense true before. Not only every single run, but even all the readings
> in every run, with a few exceptions, show the effect. As an interesting
> sort of check, the apparatus that had previously given a negative result
> was set up again; with the counters used as they were before, at lower
> voltages, the results were negative as before, but with high voltages on
> the counter, high enough to ruin the point within an hour or two, the

Table 2.2 *Experimental results of Chase (15 September 1930), relative counts*[a]

At 0°	At 90°	At 180°	At 270°	Weight
1.000	0.972	1.009	1.024	1
1.000	0.975	1.075	1.075	1
1.000	0.997	0.986	1.005	1
1.000	0.990	0.986	1.015	1
1.000	0.988	1.000	1.008	1
1.000	0.994	0.976	1.010	1
1.000	1.034	1.041	1.044	1
1.000		0.950		4
	1.000		1.030	3
	1.000		1.040	3
		1.000	1.020	2
1.000		0.933		1
	1.000		1.030	2
		1.000	0.969	2
1.000			1.003	1
1.000		1.037		2
	1.000	0.933		2
1.000 ± 0.003	0.993 ± 0.003	0.985 ± 0.003	1.021 ± 0.003	Weighted means

[a] Experimental error = 1%.

effect was very like to appear. Making no changes except in the voltage on the counter, the effect could be accentuated or suppressed.[24]

In this experiment, Chase also obtained a ratio for (count at 180°)/(count at 0°) = 0.985 ± 0.004. This time he did attribute it to a Mott scattering effect.

We are now in a position to discuss the meaning of these results. If we, in fact, accept the results of Cox and associates and Chase's last results as valid, we can conclude that the experiments give evidence for longitudinal polarization of electrons from β decay and thus for parity nonconservation in the weak interactions, as we discussed earlier.

The first question to consider is whether or not the results are statistically significant. We recall that Cox and associates obtained an asymmetry of 0.91 ± 0.01, which would normally be regarded as fairly strong statistical evidence. However, if we look at the data presented in Table 2.1, we see that there are wide fluctuations in the results and that the asymmetry is really determined by a few low results. Cox's paper also cast some doubt on the statistical reliability or validity of the result by stating that, "It must be admitted that the probable error in many cases is reckoned from too few values."[25] Despite these reservations, one should still regard these results as suggestive, at the very least. These arguments do not apply to Chase's results. He obtained an asymmetry of 0.973 ± 0.004, which is clearly statistically significant. As we saw earlier, Chase's results are also very consistent, with the asymmetry appearing not only in almost every run but also in almost every reading. We note here that seven of the ten runs that include a 90° result are within 1 S.D. of the average value, two are within 2 S.D., and only one result is 5 S.D. away. Thus, we would conclude, at least on statistical grounds, that the asymmetry was established.

There remains the possibility of systematic effects that might have influenced the result, some of which were mentioned earlier. We consider first the experiment of Cox and associates. The first problem is the wide fluctuation in the behavior of the Geiger counters. This behavior, in which some of the counters were sensitive only to slow electrons, would, in fact, lessen the effect, because the polarization effects are larger for fast electrons. There is also the possibility of misalignment of the source, the counter, or the targets. The experimenters reported that these items were removed and replaced repeatedly during the experiment and stated that "It seems unlikely that the accidental dislocations could be preponderantly in one direction as are the observations."[26] They also discussed the possibility that a stray magnetic field might have oriented the electron spins or that the electron had been polarized by passing through other material in the apparatus. They pointed out not only that such was unlikely (and, in the case of the material, had been ruled out in an auxiliary experiment) but also that it would have required "a polarity in the electron as definite as that required to explain the observations as due to double scattering."[27] There seems to be no reason not to agree with their statement that "acceptance [of the explanations] would offer greater difficulties in accounting for the discrepancies among the different results than would the acceptance of the hypothesis that we have here a true polarization due to the double scattering of asymmetrical electrons."[28] Similar arguments would also apply to possible systematic effects in Chase's experiment. This issue will be discussed in more detail in Chapter 7.

There are still several problems that can be examined only with the insight provided by later theoretical and experimental work.[29] We recall the earlier discussion of the double scattering of a longitudinally polarized electron beam. The relativistic electrons should first be scattered by a thick, low-Z target and then by a thin, high-Z target, and this will give rise to a 90°–270° asymmetry. This experiment was done in the 1950s by several groups, all of whom observed definite asymmetries of 6 percent and up.

The case involving the experiments of Cox and associates and Chase is somewhat different, and Grodzins has again provided an excellent analysis:

In the experiment of Cox et al. and Chase, the beta rays were reflected twice through 90° by thick lead and gold scatterers. The arguments above are then modified in two ways. First, the beam at $\bar{n}_2$ [Figure 2.2] will be sightly depolarized due to plural scattering[30] in the first target (a small effect as Alihanov et al. have shown). Second, the observed asymmetries should be much smaller than the recent data quoted above, since the multiple scattering in the thick second scatterer is spin independent. However, these same recent experiments argue strongly that a measurable effect should still be observed. In the three previously quoted experiments the second scatterer was so thick for the angle of scattering used, that the observed effect arose principally from plural scattering. Thus, large asymmetries are observed in the plural scattering of transversely polarized electrons. Since it is well known that plural scattering is important in scattering from thick targets, a measurable effect should be observed in the Cox and Chase experiments.[31]

There is one remaining and quite puzzling problem connected with the experimental results of Cox and Chase. In his earliest work, Grodzins concluded that these two experiments did indeed show 90°–270° asymmetry and thus could have given evidence for parity nonconservation. In a later publication,[32] Grodzins pointed out that his earlier analysis was incorrect, because both experiments found fewer counts at 90° than at 270°, whereas the modern theory predicts more counts at 90°, and thus both Chase and Cox had an effect of the wrong sign. Their error was confirmed in an experiment performed by Sidney Altman, a student working for Grodzins. Our own theoretical analysis, as well as careful comparison between the results of later experiments (Chapter 1, note

57) and those of Cox and Chase, confirms that the sign of the asymmetry obtained by Chase and Cox is indeed wrong. Grodzins, however, concluded that although the published sign of the asymmetry was incorrect, Cox and Chase had carried out correct experiments:

It has long been my view that Chase and Cox did correct experiments, but that between the investigation and the write-up the sign got changed. . . . Did Cox mislabel his angles? Did he use a right-handed coordinate system instead of the left-handed one shown in his figure? If, as I suspect, he did make some such slip then the error would undoubtedly have been retained in subsequent papers. Such errors are neither difficult to make nor particularly rare. Many a researcher and at least one former historian of science have erred similarly.[33]

I might point out that my own initial analysis was also wrong, in the same way as that of Grodzins. I also believed that the asymmetry reported by Chase and Cox was correct. We do, however, have more evidence on this point. Cox was unaware of Grodzins's later analysis, and so even though his own reminiscences are included in the same volume, he did not have a chance to respond to it. His own recollections of the problem are as follows:

I was quite surprised many years later when Lee Grodzins credited McIlwraith, Kurrelmeyer and myself with having been the first to observe parity violation. I was equally surprised; and naturally disappointed when he wrote in a later article that the asymmetry in the double scattering of β-rays, as described in our paper, was in the direction opposite to that predicted by the theory and that predicted by Yang and Lee. . . . I did not know, before the articles were printed, of the contradiction between the asymmetry predicted by the theory and that reported by McIlwraith, Kurrelmeyer, and myself, and by Chase.

Grodzins in his article expressed the opinion that we (or I should say I, since I think our paper as published was mainly written by me) made a slip between the experimental observations and its published description. He supposes that the asymmetry we found was actually in the sense the theory predicts but that, in describing the experiment, I accidentally reversed it. At first sight, at least, this seems unlikely. But the alternative explanation, which assumes a persistent instrumental asymmetry, also seems unlikely when I consider how often we removed the Geiger counter to change electrodes (as was necessary in the early short-lived type of counter which McIlwraith, Kurrelmeyer and I used) and when

I remember also other changes which Chase made in the very different equipment with which he replaced ours.

I have thought about the matter off and on for a long time without coming to any conclusion either way.[34]

Although Professor Cox draws no conclusions, I find his argument against a persistent instrumental asymmetry, both in his letter and in the published paper, convincing. The correctness of the results is not, however, important for our further discussions. Because, as we shall see, no one ever performed a similar experiment or realized the significance of the results, their accuracy was never really questioned, although some later work did attribute those results to instrumental effects. Both experiments showed the velocity dependence of the polarization that is both predicted by modern theory and observed. In view of the foregoing argument and discussions of statistical validity and systematic errors, we must agree with Cox's conclusion with regard to his own experiments and those of Chase:

It appears now, in retrospect, that our experiments and those of Chase were the first to show evidence for parity nonconservation in weak interactions.[35]

That was not, however, the reaction of the 1930 physics community. Although those results were mentioned occasionally as an anomaly, in the literature on electron scattering there was absolutely no recognition either by the authors or by anyone else of their significance for the question of parity conservation. Kurrelmeyer stated that "As to our understanding of parity, it was nearly nil. Even the term had not been coined in 1927,[36] and remember, this experiment was planned in 1925, and none of us were theoreticians."[37] Cox, in discussing the reaction of the physics community, stated, "I should say that the experiments were widely ignored,"[38] and he added that "our work was, prior to 1957, generally unaccepted, disbelieved, and poorly understood."[39] As we shall see in detail in the next section, Cox's appraisal was not entirely accurate.

2.2   THE REASONS WHY NOT

In this section we shall examine the question why these particular experiments were almost completely ignored by the physics community. The standard textbook explanation for this is that the experiments were redone with electrons from thermionic sources, rather than from β-decay sources, that do not show the effect, so that it was dismissed:[40] "... as a cure the beta decay electrons were replaced with those from a hot filament, the effect disappeared and everybody was satisfied."[41] Although there is an element of truth to this, we shall see later that it is by no means an accurate explanation.

Cox's own recollections provide a useful starting point:

As to the reaction of other physicists to the experiment of McIlwraith, Kurrelmeyer, and myself, (and also to that of Chase on the same subject) I should say that the experiments were widely ignored. I think, for several reasons, that this was to be expected. Spectroscopy was still the dominant interest in experimental physics. We were all, at the time, young and unknown. Our reported results neither confirmed nor disproved any theory which was a subject of acute interest at the time.[42]

There was no specific theoretical context, at the time, into which to place these early experiments, in contrast to the situation in 1957 when the explicit theoretical predictions of Lee and Yang were published. Cox himself supports this view:

During the nearly thirty years which passed between our experiments and those of Wu, Garwin, and Telegdi, many doubts were expressed about our observation. These doubts can be easily understood when one considers the theoretical models which prevailed before Lee and Yang. Our work was, prior to 1957, generally unaccepted, disbelieved, and poorly understood. Only by viewing it from the new theoretical framework and experimental observations of the late 50s, could our results be comprehended.[43]

We can understand why the early experiments were not regarded as "crucial," because of the lack of theoretical predictions. What is still puzzling is why the perceived anomaly in the results did not act as a stimulus for further work, both experimental and theoretical, in the same way as the θ–τ puzzle did in the 1950s, and why these results were ultimately ignored.

Before proceeding with detailed study of the reasons for this, it is worth examining, at least briefly, Cox's other points regarding fashions in physics and the prestige of authors. Although it seems beyond question that the prestige of an author, based on prior achievements, certainly influences the kind of reception that a paper receives, it seems difficult to regard that as a major factor in the treatment of these papers. As we shall see later, the results were treated seriously, at least for a time. We should also remember that the late 1920s were years of enormous excitement and growth in physics, with the rise of quantum mechanics, and that scientists like Heisenberg, de Broglie, and others who contributed so much to that development were all relatively young men at the time. Samuel Goudsmit, another of those pioneers and co-inventor of electron spin, stated this amusingly in his reminiscences regarding his own work. He quoted the biochemist Erwin Chargaff, who disparaged some members of his own field: "that in our days such pygmies throw such giant shadows only shows how late in the day it has become."[44] Goudsmit continued: "What Chargaff overlooked is that pygmies also throw large shadows at dawn. This could be applied to me and several others in the 1920s, the dawn of the new physics."[45] It is clear that Dr. Goudsmit was excessively modest about his own contribution, but this does indicate something of the spirit of the time. We might also note that at the time of publication of their Nobel Prize-winning article, Lee and Yang, although well known, were only twenty-nine and thirty-four years old, respectively.

Cox also indicated that part of the reason for the lack of interest in his work was that spectroscopy was considered a more important or more interesting area of experimental physics. Although a detailed study of fashions in physics research is beyond the scope of this book, I note briefly that Cox's view seems correct. It was primarily in spectroscopy that the predictions of the new quantum theory were being tested. Even with regard to electron spin, the major evidence was taken from spectroscopy. H. A. Tolhoek, in his review paper on electron polarization, published in 1956:

In spite of its intrinsic interest, this field of research [electron polarization] has always been somewhat out of the flow of the main efforts in experimental research, probably because physicists had already gained a firm belief in quantum mechanics of the spinning electron on the basis of other experimental data.[46]

I wish to suggest, however, that the major reason for the neglect of these results stems from the fact that they became lost in the struggle of scientists to corroborate the predictions of Mott that there should be forward–backward (0°–180°) asymmetry in the double scattering of an electron with spin, as well as in the general problem of electron scattering from nuclei (Mott scattering). As we shall see later, this problem was of concern until the 1940s and even into the 1950s, with difficulties in the consistency of experimental results and with subtle and unforeseen effects in the scattering of the electrons. These experimental difficulties also led to various theoretical attempts to modify Mott's theory to explain the absence of the predicted effects. It is worth examining that history in detail, not only because it provides an explanation for the neglect of the experiments of Cox and his collaborators and of Chase but also because it gives an insight into how the physics community deals with an apparent discrepancy between experimental observations and a theory that is believed to be true for other experimental and theoretical reasons. That history also indicates that certain technological advances, namely, the development of electron accelerators with energies of hundreds of kilovolts, hindered an understanding of these early results, as contrasted with the situation in the 1950s, when advancing technology made an understanding of the anomalous θ–τ puzzle more important.

As indicated earlier, the experiments on double scattering of electrons to demonstrate electron polarization, which were done before Mott's theoretical calculation in 1929, were based on analogies to optical scattering. In this case, the first scattering acts as a polarizer, and the second as analyzer. With the exception of the experiment of Cox and associates, which we have discussed at length, none of these early experiments gave any evidence of electron polarization.

Wolf[47] attempted to demonstrate electron polarization by aligning electron spins by passing them through a magnetic field and observing the polarization by scattering from a metal surface, but he observed no effect. In fact, in a later argument, Bohr showed theoretically that this experiment could not give any positive results.[48] Rupp did an experiment that was quite similar to that of Cox (i.e., two 90° scatters), although at low energy (up to 380 V), and observed no effect.[49] When he changed the first scattering angle to 12°–14°, he observed a difference between the 90° results and the results at 0° and 180° that he attributed to an instrumental effect, although he was not certain of his explanation. A similar small-angle experiment (10°–30°) was done by Joffé and Arsenieva,[50] who reported that their null result was consistent with preliminary results of Davisson and Germer[51] and also with a theoretical calculation by Frenkel,[52] who claimed to have shown the impossibility of producing an electron polarization by reflection from a metal surface.

In 1929, Davisson and Germer[53] published a detailed report of their work. Their experiment involved two 90° scatters from nickel crystals at energies of 10–200 V. They concluded: "Our observation is that electron waves are not polarized by reflection." They were, however, the first authors to discuss the work of Cox and his collaborators. They noted the general similarity of the experimental arrangements,

But in other respects the experiments differ. The electrons constituting the primary beam are β-rays from a sample of radium, the reflections are plates of polycrystalline gold, and the collector is a point-discharge electron-counter. The authors report that the shielding between the electron source and the counter was inadequate to suppress entirely an effect due to the gamma-radiation, and further that rapid changes in the characteristics of the discharge point made it difficult to obtain consistent data. The results which they publish are ratios of the current received by the collector in one of the "parallel" [0° or 180°] positions to that received in one or the other of the "transverse" positions [90° or 270°], and the ratios of the currents received in the two "transverse" positions. The values found in the first of these ratios depart from unity by much more than the probable error, and show a bias in favor of polarization. The authors do not point this out, however, but lay emphasis instead upon a rather slight departure from unity of the values obtained for the second ratio – that of the currents in the two transverse directions.[54]

Davisson and Germer were not questioning the correctness of Cox's results, but rather were changing the emphasis to underline the differences between the counts at 0° and those at 90° or 270°, which were more consistent with their expectations from the optical analogy.

The situation changed in 1929 with the publication of Mott's theoretical calculation of the double scattering of electrons.[55] Mott's calculation was based directly on Dirac's relativistic electron theory[56] and made specific theoretical predictions about the asymmetry to be observed in the double-scattering experiment (see the discussion in Section 2.1). He predicted that the number

of electrons detected when the source (Figure 2.1) was at 180°
would be greater than when the source was at 0°. Mott specified
the conditions under which this asymmetry should be observed,
namely, single, large-angle scattering of high-velocity electrons
from a high-Z nucleus. He also gave a formula for the magnitude
of the effect expected, although he subsequently modified that
and provided accurate numerical results in subsequent papers
published in 1931[57] and 1932.[58] In his 1929 paper, Mott particularly
noted that his theory did not predict any asymmetry between the
90° and 270° directions:

The greatest asymmetry, therefore, will be found in the directions
$TM(180°)$, $TM'(0°)$, in the plane of the paper. In the plane through T_1T_2
(the line joining the two targets) perpendicular to the plane of the paper;
the scattering is symmetrical about T_1T_2. It was in this plane that asymmetry was looked for by Cox and Kurrelmeyer, and the asymmetry
found by them must be due to some other cause.[59]

Mott was remarking here only that his theory that predicts that
the first scattering produces a polarization that results in 0°–180°
asymmetry in the second scattering did not apply to Cox's results.
He did not seem to be questioning the validity of the results.

Subsequent experimental work during the 1930s took on a different character following Mott's researches, because there were
then explicit theoretical predictions, based on an accepted theory
(namely, Dirac's electron theory), with which to compare the
experimental results. The experimental situation was confused at
best. We find some experimenters finding the predicted results,
others doing similar experiments and obtaining null results, and
some experimenters finding positive results at one time but not
at others. In general, however, the trend in the experimental
results was in disagreement with Mott's calculation.[60] This discrepancy between theory and experiment led not only to further
experimental work but also to many attempts by theoretical physicists to give reasons for the apparent absence of the polarization
effects predicted by Mott. As we shall see later, these attempts
were quite unsuccessful.

The earliest publication that discussed Mott's work seems to
have been Chase's experiment published in 1929:

A theory of the experiment has just been published by Mott, who considers the double scattering of Dirac electrons by atomic nuclei. This

theory, while not as yet entirely complete, predicts a difference between the counts at 0° and 180°. The data just given show that more electrons were counted at 0° than 180°, but this has been ascribed to the somewhat freer path through which the electrons have in the 0° position.[61]

In his later paper, however, Chase does seem to claim to have observed the effect:

A previous experiment of the writer showed an effect of this nature, more electrons being counted in the 0° position than in the opposite direction by 4%. Nothing was claimed for this result, because the targets were so close together that it seemed plausible that the electrons would find it easier to get through the apparatus in the 0° position. The targets are now farther apart, since fewer electrons per second are permissible with the increased sensitivity of the electroscope, and beam divergence is much less than before. The previous objection is no longer valid.[62]

Rupp also continued his work on electron scattering in the light of Mott's theory. His experiment[63] examined the double scattering of electrons at grazing incidence (approximately ⅓°). He observed an increase in the polarization effect with increasing electron energy, as predicted by Mott, and obtained a difference of 11.8 percent from gold at an energy of 80 keV. Because he also observed no effect for scattering from beryllium, again in agreement with Mott, Rupp claimed to have corroborated Mott's theory. As Chase[64] correctly pointed out, however, Mott's theory applied only to large-angle scattering. Rupp subsequently corrected his numerical results,[65] but without modifying his conclusion. His data also indicate equal scattering at 90° and 270°. In a theoretical paper on electron polarization by Fues and Hellmann,[66] Rupp's results were regarded as a corroboration of Mott's theory. Fues and Hellmann noted Chase's 1929 results, mentioned earlier, in which the interpretation was uncertain, and were also aware of his later results.

Kirchner[67] performed an experiment similar to that of Rupp, and although he claimed to have observed some asymmetries, he gave no quantitative results. He remarked that these asymmetries were no larger than 10 percent and stated that the experimental data were insufficient to rule out Mott's theory. G. P. Thomson also attempted to reproduce Rupp's results by small-angle scattering through two gold foils. He obtained a negative result and wrote that "the experiment is in agreement with the view that the detection of polarization by such means is only possible with large angles of scattering."[68] This view was supported by Halpern[69] in his theoretical interpretation of Rupp's results. He stated, correctly, that Mott's theory did not apply to Rupp's experiment and offered his own theory for scattering from a two-dimensional grating as a substitute, although he did not give any quantitative results. Rupp continued experimenting, this time with an apparatus that was closer to that required by Mott's theory.[70] He first scattered the electrons at 90° off a gold reflector and then passed the beam through a gold foil and photographed the resulting pattern. He again observed an asymmetry. In a subsequent experiment, Rupp and Szilard[71] further corroborated these results and obtained some additional results using magnetic fields to precess the electron spins, in agreement with theory.

In 1931, Mott published a short note[72] that modified his previous calculations slightly and also gave numerical results. He noted briefly that Dymond[73] had performed the experiment and found an effect that was five times too small (later we shall discuss Dymond's complete report on his experiment):

The asymmetry at 70 kV is thus about five times as much as that found by Dymond. It is difficult to explain why this should be so. Multiple scattering would reduce the polarization observed, but there should not be much multiple scattering with the foils used. It is improbable also that the Dirac theory of the electron should give a wrong result when applied to the scattering by a Coulomb field, since the results for the energy levels of an electron in the same field are known to be correct.[74]

The experimental situation became further confused by Rupp's continuing experiments. He repeated his experiment on small-angle scattering from foils of different materials and again observed asymmetries, although this time he noted that Mott's theory did not apply.[75] His next experiment[76] was a reasonable approximation to the conditions required by Mott's theory. The electrons were first scattered at 90° by a gold foil and scattered again at 90° by a gold wire. He obtained clear 0°–180° asymmetry of 3–4 percent at an electron energy of 130 keV and 9–10 percent at 250 keV, with an error of 1–2 percent. These results were in quantitative disagreement with Mott's numerical predictions of 15.5 percent at 127 keV and 14 percent at 204 keV.[77] This discrepancy was corroborated by more accurate measurements done somewhat later[78] that gave 3.8 ± 0.5 percent at 130 keV and 9.6 ± 0.5 percent at 250 keV. Rupp proposed that perhaps the polarization of the nucleus itself might be important and that this might explain the discrepancy, although he clearly regarded the lack of quantitative agreement as a serious problem.

Mott further modified and corrected his calculations and presented numerical results in a paper published in 1932.[79] Although he did not mention any specific experimental results, he seemed aware of the discrepancy between theory and experiment. He emphasized the conditions necessary for his theory to hold (i.e., single large-angle nuclear scattering of high speed electrons from a high-Z nucleus). After presenting some other possible methods for producing polarized electrons, he noted that

These methods of producing polarized electrons are naturally quite beyond the range of present experimental techniques. Nevertheless, the theoretical existence of these models provides some evidence that electron beams can be polarized; for if they were to fail to produce a polarized beam a far more drastic revision of present-day quantum mechanics would be required than would be necessary if the double scattering experiment should give a negative result.[80]

Although Mott clearly believed that failure to observe these new predicted effects would be more serious, he was indicating that failure of the double-scattering experiment would also require a modification of quantum mechanics.

By that time the experimental situation was so confused that all authors felt obliged to review the past history in some detail. Thibaud and associates[81] divided previous experiments into two groups: The first included those that gave either a negative result or a positive result that was within experimental error of a null result. Included in this group were the experiments of Cox and associates and Chase's 1929 experiment, which gave a null result. They made no mention of Chase's later positive result, which seems surprising, because they did refer to other later papers. The second group consisted of Rupp's experiments, and those were regarded by Thibaud and his collaborators as giving a positive result. Their own experiment, however, did not satisfy the conditions for Mott scattering. They performed a single small-angle scattering, followed by diffraction through a thin foil, and observed no effect.

Langstroth[82] performed a double 90° scatter from thick tungsten targets at 10 keV and observed no effect. Thus, he was in agreement with Mott's calculation, which predicted no polarization at such a low energy. His review of other work shows clearly the confusing experimental situation:

With very much higher energies, however, an asymmetry in the intensity distribution of the secondary scattering appears to exist, although the evidence is somewhat contradictory and incomplete . . . .
   Experimentally asymmetric distributions have been observed with fast electrons for both small and large scattering angles. The reported effects are of a quite different kind; varying apparently with the method of production. With small angle scattering from solid targets an effect has been reported which is not found when the scattering occurs through foils. Moreover, a still different effect is observed when a combination of solid target and foil is used. An additional type of asymmetry (in the 90–270° plane) was first observed by Cox et al., but was later attributed to instrumental causes. More recently, however, Chase reported that this type of asymmetry does exist. Finally, a small asymmetry (about 2%) of the type predicted by the specialized theory [Mott] has been observed by Dymond in electron scattering from thin foils.[83]

The results of Chase were regarded as valid, but they were only one set in a collection of confusing results.

After a somewhat more detailed account of the earlier experiments, Langstroth tried to offer some reasons why the Mott theory did not apply: "In view of the fact that practical conditions may be immensely more complicated than those of Mott's theory, it is not surprising that it does not furnish a guide, even in a qualitative way, to all of the above experiments. This may be due to (a) the fact that a large proportion of the beam scattered from a thick target consists of electrons which have undergone more than one collision, (b) the insufficiency of the theoretical model, (c) the inclusion of extraneous effects in the experimental results."[84]

Langstroth discussed these possible effects in detail and mentioned other theoretical efforts, namely, those of Frenkel and Halpern, without reaching any conclusion.

The difficulty of offering a consistent interpretation of the experimental results was increased with the publication of Dymond's detailed account[85] of the experiment mentioned earlier. Dymond's apparatus came quite close to the conditions required by

Mott. He used double 90° scattering by thin gold foils at an energy of 70 keV. In order to guard against instrumental or geometric asymmetries, he also took data at 20 keV, where the predicted asymmetry is very small, and subtracted the results at 20 keV from those at 70 keV to determine the polarization. As a further check, he substituted one aluminum foil for a gold foil. Mott's theory predicted that there should be almost no polarization from aluminum, and Dymond obtained a result of 0.11 ± 0.10 percent polarization, consistent with zero. Dymond's results for the scattering off two gold foils was 1.70 ± 0.33 percent. His data showed a fair degree of consistency and were in the direction predicted by the theory (i.e., more scattering at 180°). He was greatly concerned, however, that Mott's theory predicted a polarization of approximately 10 percent at that energy and that his result was six times too small. He offered a possible explanation for his low result by considering plural scattering, in which several reasonably large angle scatters add up to a 90° scattering. He noted, however, that "Although it is not possible to estimate the influence of plural scattering, it does not seem likely that it can be called upon to bridge the gulf between the experimental and theoretical values."[86] This is a point we shall return to later. Dymond also believed that the theoretical calculations had been done correctly:

On the theoretical side, no hopes are entertained that a lower value may be the correct one. Mott points out that the Dirac relativistic wave equation leads to the correct solution of the energy states in a Coulomb field, and it is not likely to fail in the problem under discussion.[87]

After a brief discussion of Rupp's results and the difference between them and the results of G. P. Thomson and Kirchner, Dymond concluded that "the final conclusion on the present evidence is that there is no concordance with theoretical expectations."[88]

Dymond, interestingly, performed yet another check on his apparatus by measuring the scattering at 90° and 270°, where he expected no asymmetry. To his evident surprise he found a polarization of −1.75 ± 0.98 percent, which indicated more scattering at 270°. He noted, however, that the data were quite inconsistent, with the sign of the effect changing from run to run. He again used a single aluminum foil and obtained a result of + 1.97 ± 0.84 percent. If one regards the aluminum data as giving the geometric or instrumental asymmetry, then one should subtract the aluminum data from the gold data to obtain the correct asymmetry. That gives a result of − 3.7 ± 1.3 percent. Although Dymond did not perform this calculation, he was obviously concerned about these results:

As will be seen . . . the mean difference is only slightly larger than the probable error, while the asymmetry when one foil is aluminum is twice the probable error. The results shown . . . would emphatically lead to the conclusion that there is no real asymmetry between the two azimuths, in accordance with the theoretical predictions. The very much greater concordance of the results in the 0°–180° plane, where there is no single instance of a reversed sign, leads me to believe that here the asymmetry has a real significance. Nevertheless, the behavior of the apparatus in the 90°–270° plane is disturbing, as no reason can be given for the results found.[89]

What is most surprising here is that Dymond made no mention at all of the similar results of Cox and associates or of Chase's 1930 result. He was obviously aware of Chase's paper, because he noted early in his own work, in reference to the 0°–180° effect, that "Chase has carried out experiments with electrons of somewhat higher velocities (β-rays), but finds no significant polarization."[90]

Our previous analysis of double scattering (Section 2.1) shows that there should be no 90°–270° asymmetry for thermionic electrons, which are initially unpolarized. The effect should be observed only by using the longitudinally polarized electrons from β decay. Thus, in retrospect, Dymond was right in doubting the correctness of his results.

Both Thomson and Rupp continued their experiments, with Thomson again failing to find any asymmetry in a 90° scattering off gold, followed by diffraction.[91] Rupp, on the other hand, repeated his double 90° scattering experiment, this time using thallium vapor rather than gold foils.[92] He again reported clear evidence of the asymmetry predicted by Mott. Although his results were still below the theoretical predictions, they were in closer agreement than the previous results with gold. It seems likely that Rupp used the thallium vapor to test his earlier hypothesis that polarization of the nucleus was the reason for the discrepancy. He noted, correctly, that multiple scattering would be larger in the gold foil and would depress the gold values.

At approximately the same time, Sauter[93] recomputed Mott's calculation by another method and obtained identical results. He also considered the question whether or not screening of the nucleus by atomic electrons might cancel the polarization effects and found that, except for slow electrons, where no polarization was expected, this was not the case.

The situation became even more complicated in early 1934, when Dymond published a full repudiation of his earlier results.[94] In continuing his experiments, he found that he obtained results that were of opposite sign and increased with increasing electron energy. On examining his apparatus, he discovered that there was considerable instrumental asymmetry:

It had always been assumed previously that any such asymmetry would be independent of electron velocity; and so would be eliminated by the method of combining observations at a high and low voltage. It seems from these experiments that this is not always a justifiable procedure.

In order to test this conclusion, the error in position of the first film was somewhat over corrected. The value of δ [the asymmetry] now changed sign . . . . On readjusting the height of the film as accurately as possible, the values in [the table] were obtained.

It is therefore apparent that by accurate adjustment the value of δ can be reduced to zero, and by deliberate maladjustment may be made to assume either positive or negative values. There is no evidence of a true polarization larger than 1% in these measurements, and there is no reason to believe that the small positive result $2\delta = 1.7\%$ at 70 keV, found in [his previous paper] is not due to a slight error of adjustment as just described.[95]

Dymond then considered possible reasons for the discrepancy between theory and experiment, namely, inelastic, stray, and plural scattering, and nuclear screening. He rejected all of these, citing Sauter's work, and concluded that

We are driven to the conclusion that the theoretical results are wrong. There is no reason to believe that the work of Mott is incorrect; . . . It seems not improbable, therefore, that the divergence of theory from experiment has a more deep-seated cause, and that the Dirac wave equation needs modification in order to account successfully for the absence of polarization.[96]

Dymond also noted the conflicting results of Rupp and Thomson, and though he obviously had doubts about Rupp's work, he drew no conclusions.

Thomson also published a comprehensive review[97] of the field, along with his own detailed results, both on a single 90° scattering from thin foils, followed by diffraction through a foil, and on double 90° scattering. In the former experiment he found no effect of the type previously found by Rupp. In the latter, he found a ratio of counts at 0° to counts at 180° of 0.996 ± 0.01 at an energy of 100 keV, whereas Mott's calculations required a result of 1.15. Thomson remarked that his diffraction results were in agreement with theoretical expectations, where no polarization is predicted at small angles, but he noted that Rupp's effects all appeared at higher energies. He suggested that Rupp may have had a non-uniform background of scattered electrons, but he was concerned about Rupp's results using magnetic fields. In reviewing the double-scattering experiments, Thomson pointed out that all of them, those of Langstroth, Dymond, and Rupp, previously mentioned, made use of a comparative method, comparing their data to those obtained either at low energies or by scattering from low-Z nuclei, where no polarization effects were expected. He stated that because the results of most experiments were in disagreement with Mott's predictions, then use of the theory to analyze the data was itself open to question: "... if Mott's theory is wrong in some respects it may be wrong in others, and the assumption that aluminum gives much less polarization than gold becomes doubtful."[98] Thomson's conclusion about his own double-scattering experiment is quite revealing with regard to the problem of the discrepancy between theory and experiment:

We have seen that there is no polarization of the kind required by Mott's theory within the limits of the probable error, which we may take as ±1%. Certainly the 14% effect required by the theory does not exist. Most of Mott's work has been, I understand, checked independently; but unless there is some error in the calculations, we are driven to the conclusion that Dirac's theory cannot be applied to the problem of scattering by heavy atoms. It seems very unlikely that the presence of the outer electrons can make much difference to the scattering, which must be almost entirely nuclear .... All the conditions of Mott's calculations were fulfilled in these experiments, but it seems possible that Dirac's theory doesn't hold in the intense field close to a gold nucleus where, of course, most of the scattering takes place. It is probable that an electron with sufficient energy can cause a breakdown in the medium in such a region and produce a pair of positive and negative electrons, as quanta are known to do. While the electrons I used would certainly not be able to do this, it is not unreasonable to suppose that the Dirac equation ceases to hold exactly for energies much below those required to cause a breakdown. It is, however, surprising that the effect should be just to cancel the polarization, and one is inclined to surmise that the polarization of a free electron may be among the unobservables.[99]

Thomson did make a brief reference to the work of Cox and to that of Chase, although again only to the 1929 publication:

The first experiments with fast electrons were made with β-rays by Cox, McIlwraith, and Kurrelmeyer, and continued by Chase. The effect found, about 4%, in the opposite direction to that predicted by Mott, was considered to be instrumental.[100]

Thomson seems to have been the first to question the validity of those results, but it is again surprising that no mention was made of Chase's later work, which was much more reliable. The lack of mention of the 90°–270° asymmetry is less surprising, because that cast no light on the problem he was considering.

Rupp,[101] at that same time, continued to refine his double-scattering experiment, with particular emphasis on the effects of both longitudinal and transverse magnetic fields. He regarded that experiment as a test of the relativistic formulas used to calculate the magnetic field effects. He did not mention Mott's theoretical work at all, perhaps because of the confused situation. Myers and associates[102] attempted to corroborate the earlier results of Rupp, and Rupp and Szilard, in a double-diffraction experiment. They found no evidence of asymmetry and discussed a possible cause for the positive results of Rupp's experiments. They summarized the experimental situation as follows:

Decided evidence of polarization is reported in all the experiments of Rupp, and Dymond[103] obtained a positive result, which, though small was significantly larger than the estimated error of observation. In the other experiments cited, the results were negative, or, as in those of Cox, McIlwraith, and Kurrelmeyer, and Chase, rather inconclusive, because of experimental difficulties.[104]

At that time, Cox and his collaborators were expressing doubts about their earlier results.

The general trend of the experimental results, which was in disagreement with Mott's theory, led to several theoretical attempts either to modify that theory or to present a new one that would be in agreement with the data. Hellmann,[105] in 1935, offered a theory of electron scattering from a layered homogeneous potential and concluded that no polarization was to be expected from such scattering. He noted that this was in contrast to the results of Mott's work, but regarded his own calculation as more realistic, which was not the case. Halpern and Schwinger[106] modified the Coulomb potential by addition of a repulsive term of the form $V = b/r^5$ and found that it "annihilates the polarization effect completely." Winter[107] also obtained a negative result for double scattering of monochromatic Dirac plane waves from a sphere of constant potential and remarked that that explained the negative results of the then recent experiments. We note here the ad hoc nature of these theoretical discussions. They do not seem to have been regarded as solving the problem of Mott scattering.

In 1937, Richter[108] published what he regarded as the definitive experiment on double scattering of electrons. He claimed to have satisfied the conditions of Mott's theory exactly and found no effect. He concluded that

Despite all the favorable conditions of the experiment, however, no sign of the Mott effect could be observed. *With this experimental finding, Mott's theory of the double scattering of electrons from the atomic nucleus can no longer be maintained* [emphasis in original]. It cannot be decided here how much Dirac's theory of electron spin, which is at the basis of Mott's theory, and its other applications are implicated through the denial of Mott's theory.[109]

Richter included the experiment of Cox and associates in his survey of previous experiments and remarked that they did observe an asymmetry. He also noted that Chase's 1929 paper reported an asymmetry, which was zero within experimental error. No mention was made of Chase's 1930 results or the 90°–270° asymmetry, showing that this effect had been lost during consideration of the problem of the Mott theory.

In 1938, Champion[110] reviewed the entire subject of electron scattering and noted that there were discrepancies between Mott's theory of single scattering and the experimental results:

This theoretical work is, of course, based on Dirac's relativistic wave-equation for an electron, and it appears to be quite impossible that any slight modification of existing theory can be made to account for a divergence of the magnitude found experimentally . . . we should mention here that Richter has examined the double scattering of electrons of energies up to 0.12 mV at 90° by gold foils. The results are in complete disagreement with the theory, no polarization can be observed, whereas 16% of polarization is to be expected from the theory of Mott, based on Dirac's relativistic wave-equation.[111]

Champion's view that there was a definite discrepancy between theory and experiment was confirmed in a 1939 paper by Rose and Bethe entitled "On the Absence of Polarization in Electron Scattering."[112] They tried various ways of solving the problem and concluded as follows:

In addition to multiple scattering we have investigated the depolarizing effect of other processes. These are: (1) inelastic scattering with change of spin of the incident electron and (2) exchange scattering in which the exchanged electrons are of opposite spin. The result of these considerations may be stated very briefly. Unfortunately, none of the effects considered produces any appreciable depolarization of the electrons and the discrepancy between theory and experiment remains – perhaps more glaring than before.[113]

Bartlett and Watson refined Mott's numerical calculations and observed that "Our polarization values are about 10% less than those of Mott at the maximum, so that the theory still predicts an effect which has not been observed."[114]

In 1939, Kikuchi reported preliminary results that seemed to be in agreement with Mott's theory:

Of course, it is quite beyond the range of experimental technique to realize perfectly all conditions from which Mott's theory is originated. In fact, the thickness of the gold foil may be too large to be applicable for the theory.[115]

Kikuchi's view was accepted by Rose[116] in another paper that attempted to modify the theory. Rose noted that because Kikuchi had used such thick targets, his results, which were in apparent agreement with the theory, were in fact much larger than the theory predicted, because multiple scattering should result in 80 percent depolarization. He suggested that Kikuchi's results might be attributable to an instrumental asymmetry similar to that of Dymond. Rose's attempt to modify the theory to eliminate the discrepancy resulted in failure, once again:

The possibility of accounting for the small polarization of electrons, as observed in the double scattering experiments and the anomalously small scattering of fast electrons ($E > 500$ keV) in heavy scattering materials, by the assumption of non-Coulombian forces near the nucleus is investigated. . . . Therefore in order to obtain an asymmetry and scattering which are not in obvious disagreement with the observations it is necessary to assume either a large range or a specialized form for the non-Coulombian forces. . . . Insofar as these possibilities do not seem plausible, it would appear that the anomalous scattering and asymmetry must be explained on grounds other than the existence of non-Coulombian forces.[117]

It is fair to say that in early 1940, the situation with respect to Mott's prediction of a polarization effect to be observed in double scattering of electrons was as follows. Although Rupp, and later Kikuchi, had observed such effects, the preponderance of evidence was that such effects had not been observed. Various theoretical attempts had been made to resolve the apparent discrepancy between theory and experiment, with no success. I argued earlier that it was in fact this problem, which seemed so crucially important to the Dirac theory of the electron, that led physicists to neglect the anomalous results of Cox and associates and those of Chase. This certainly is borne out by our examination of the literature of the 1930s. It is therefore ironic that the solution to the problem was begun in the work of Chase, Cox, and their collaborators.

In August 1940, Chase and Cox published a paper on single scattering of 50-keV electrons from aluminum.[118] They noted that previous experiments on the subject had

indicated exceptions to this equation and to the closely related prediction of an asymmetry in double scattering, and it has been suggested that they show an actual invalidity of the Dirac equations in a range of electron speeds where it is hardly to be expected on other evidence.[119]

In performing that experiment, they found a totally unexpected asymmetry for which they could offer no explanation:

It appeared in a comparison of the intensities of the beams scattered at 90° on the two sides of the foil, the side on which the beam was incident and the opposite side. Single nuclear scattering should be equally intense on both sides. But the observed scattering was consistently more intense on the side on which the beam was incident.[120]

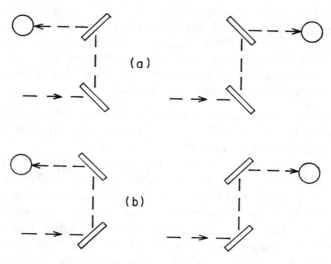

Figure 2.3. An ideal transmission experiment on double scattering (a). An ideal reflection experiment (b).

They rejected other possible scattering effects and instrumental asymmetries and ultimately found results in agreement with Mott's theory after a suitable averaging process.

This reflection–transmission asymmetry was then used in analyzing the results of a double-scattering experiment by Shull in a preliminary report,[121] and in more detail by Shull and associates.[122] In those papers they noted that the polarization observed depended on whether it was a "transmission" or a "reflection" polarization experiment (Figure 2.3). They pointed out that a much larger effect was observed in the transmission experiment:

This large asymmetry is interesting in that past theories of particle scattering by thin foils have overlooked the possibility of such an effect. Goertzel and Cox[123] [we shall discuss this paper later] show that an effect of this kind may be caused by a type of plural scattering which consists in a combination of two deflections of the same order of magnitude. . . . If plural scattering is responsible for this effect, it is possible that a polarization experiment similar to the present one might be seriously affected by it because of the depolarization which accompanies plural scattering. There is the possibility then that a reflection polarization experiment (in which only "reflected" electrons are studied) will yield an asymmetry different from a transmission experiment (in which only "transmitted" electrons are studied).... In comparing experimental results with the theoretical predictions those obtained in an ideal transmission experiment should be used, since here the theoretical requirement of single scattering is met most faithfully.[124]

They analyzed their transmission data and obtained a polarization asymmetry of 1.12 ± 0.02, in good agreement with that of 10.4 percent calculated by Bartlett and Watson.[125]

These authors also examined the previous experiments that gave no polarization effects, most notably those of Dymond and Richter, and noted that both experiments were reflection experiments, in which the polarization effect was expected to be considerably smaller. We can understand this using the theory developed by Goertzel and Cox.[126] The polarization effects are large only for scattering angles near 90°, and thus electrons scattered at 45° will not be polarized very much. We also note that the probability of Coulomb scattering is proportional to $1/\sin^4 (\theta/2)$ and thus decreases rapidly with increasing angle. In a reflection experiment (Figure 2.3), a 90° scatter can be made up of either two 45° scatters or two 135° scatters. This is called plural scattering. We neglect the latter because such scattering is very small. The first 45° scattering will result in the electron going along the foil, passing through a large amount of scattering material, and making the second 45° scattering quite likely. Because the electrons thus scattered are not polarized, they can mask the polarization expected by a single 90° scattering. In a transmission experiment, however, a 90° scattering can be made up only of a combination of 45° scattering and 135° scattering. This plural scattering will be considerably smaller than the two 45° scatters, because of the $1/\sin^4(\theta/2)$ dependence of the scattering, and thus the polarization will not be masked.

It is sad to note that after this work, which initiated the resolution of the Mott scattering problem, and in which Professors Cox and Chase played such important roles, they themselves had forgotten about their earlier scattering anomaly. Although there is a reference[127] to the earlier attempt of Cox and associates to find evidence for polarization of free electrons, no mention was made of the 90°–270° asymmetry found by them and subsequently by Chase. Although much work remained to be done,[128] primarily in improving the quantitative agreement between theory and experiment, it is clear that the major breakthrough in the problem of Mott scattering was provided by the work of Chase, Cox, and their collaborators.

We mentioned previously that a major factor in the failure of the physics community to understand the early results of Cox and associates and those of Chase was the development during the decade of the 1930s of electron accelerators with energies on the order of hundreds of kilovolts. As we have seen, there were numerous difficulties in performing experiments on Mott scattering, and the development of accelerators in which beam size, direction, and energy could be controlled precisely was seen, justifiably, as an important technological advance. With the exception of the early work of Cox and associates and Chase, who used high-energy electrons from β decay, all experiments used artificially accelerated electrons. This technological advance, however, precluded any confirmation of the anomalous results of Cox and Chase. As we saw earlier (Section 2.1), the 90°–270° asymmetry observed by them can occur only for electrons that are longitudinally polarized initially. Thermionic electrons, which are initially unpolarized, can give rise to only the 0°–180° asymmetry predicted by Mott. It is clear that no physicist of that time thought that the difference between thermionic electrons and those from nuclear β decay was of any significance. Cox himself confirmed that view:

For some years a small group of us at N.Y.U. continued experiments in the scattering and diffraction of electrons. But, as well as I can remember, most of our experiments were not with β-rays but with artificially accelerated electrons. Although the title of our first paper was "Apparent Evidence of Polarization in a Beam of β-Rays," I did not suppose, and I do not think the others did, that β-rays were polarized on emission. I thought of the targets as having the same effect on any beam of electrons at a given speed, polarizing at the first target, analyzing at the second. Consequently I did not think of the change from a radioactive source to an accelerating tube as a radical change in my field of research.[129]

2.3 DISCUSSION

It seems clear that the experiments of Cox, McIlwraith, and Kurrelmeyer and those of Chase show, at least in retrospect, nonconservation of parity. It is also true that the significance of those experiments was not recognized by anyone in the physics community at the time. At least part of the reason for this was the lack of a theoretical context (such as existed in 1956 during the work of Lee and Yang) in which to place the work. I have also argued that the reason these experimental results, which were originally considered valid and which were not predicted by any existing theory, did not lead to any further theoretical or experimental work was that they became lost in the struggle to solve the discrepancy between Mott's calculation (based on Dirac's electron theory) and the experimental results on double scattering of electrons in the 1930s. The advance of technology, in which electron beams of high intensity, good resolution, and controlled high energy became available with the development of electron accelerators, also precluded the possibility of reproducing the results of Cox and Chase, which depended on longitudinal polarization of electrons from β decay. At the time, no one realized that there was any difference between thermionic and decay electrons.

This episode also illustrates one possible reaction of the physics community to a seemingly clear discrepancy between experimental results and a well-corroborated theory. Dirac's theory was not rejected or regarded as refuted, even after many repetitions had seemed to establish the discrepancy beyond any doubt. Repetitions of the experiment continued, under similar and slightly different conditions, and various ad hoc theoretical suggestions were made to try to solve the problem, all of which were unsuccessful. As we have seen, the discrepancy was finally resolved by an experimental demonstration, followed by a theoretical explanation, of why the earlier experimental results were wrong. A discussion of the difference between this episode and that of parity violation, in 1957, when a strongly held law was refuted, will be presented in Chapter 4.

3

CP or not CP

In this chapter we shall examine the role of the experiment of Christenson, Cronin, Fitch, and Turlay[1] in refuting the concept of invariance of physical processes under CP (combined space inversion and particle–antiparticle interchange). We shall see that this experiment was a "convincing" experiment. It persuaded most of the physics community that CP was violated. There were, however, several alternative explanations offered to explain this result, and we shall also discuss the role that experiment played in refuting or eliminating these alternatives. We begin with the historical background of the experiment and then examine both the experiment itself and the reaction of the physics community to it.

3.1 HISTORICAL BACKGROUND

From the time of their discovery in 1947 by Rochester and Butler,[2] the neutral K mesons, or kaons, have exhibited odd features. One puzzle, that certain K particles of equal masses and lifetimes (called θ and τ) seemed to decay into states of opposite parity, was resolved by the discovery of parity nonconservation (see Chapter 1). An equally puzzling observation concerning the kaons was put this way by C. N. Yang: "While the[se] strange particles are produced quite abundantly (say 5% of the pions) at BeV energies and up, their decays into pions and nucleons are rather slow (10^{-10} sec). Since the time scale of pion-nucleon interactions is of the order of 10^{-23} sec, it was very puzzling how to reconcile the abundance of these objects with their longevity (10^{13} units of time scale)."[3] In 1952, A. Pais suggested that strange particles appeared in the strong interactions only in pairs, which he called "associated production."[4] When they decayed, however, they

necessarily did so one at a time, apparently through weak interactions, which occurred more slowly than strong ones. In an ad
hoc way, at least, that explained the phenomena. Pais's suggestion
was elaborated in the theory of strangeness conservation put forward by Gell-Mann and Nishijima between 1953 and 1955.[5] They
supposed that K mesons and hyperons (particles heavier than
nucleons) possessed a quantity called strangeness that was conserved in strong interactions but not in weak interactions, and so
accounted for the copious production and long decay time of
strange particles. Conservation of strangeness would forbid the
occurrence of certain strong reactions and decays otherwise allowed. No violation of this restriction had been noticed by 1956.[6]

One consequence of this scheme was that there would be two
doublets: K^0 and K^+, with strangeness $+1$, and $\overline{K}^0$ and K^- with
strangeness -1. The K^0 and $\overline{K}^0$ constitute a particle–antiparticle
pair, quite unlike the neutron and antineutron, for whereas inter-
conversion in the latter case is forbidden by baryon conservation,[7]
the K^0 and $\overline{K}^0$ can transform into one another by the weak inter-
action $K^0 \leftrightarrow \pi^+ \pi^- \leftrightarrow \overline{K}^0$, in which strangeness is not conserved.
Gell-Mann and Pais, who pointed out this distinction, suggested
that although the K^0, $\overline{K}^0$ description was appropriate for the production of these particles by the strong interaction, in which
strangeness is conserved, the weak-interaction transformation
noted earlier indicated another description that better fit the decays resulting from the strangeness-violating weak interaction.
They defined two new particles, K_1^0 and K_2^0, as the decayers:

$$|K_1^0\rangle = (|K^0\rangle + |\overline{K}^0\rangle)(1/\sqrt{2}) \tag{3.1}$$

$$|K_2^0\rangle = (|K^0\rangle - |\overline{K}^0\rangle)(1/\sqrt{2}) \tag{3.2}$$

K_1^0 and K_2^0 are superpositions of $\overline{K}^0$ and K^0, and vice versa.[8]

At that time, 1955, it was believed that all interactions remained
invariant under the operator C (charge conjugation or particle–
antiparticle interchange) and that therefore K_1^0 and K_2^0 were eigenstates of C, with eigenvalues of ± 1, respectively. Gell-Mann
and Pais further concluded that K_1^0 and K_2^0 must have different
decay modes and different lifetimes. In particular, they noted that
only one of the pair, later found to be K_1^0, could decay into two
pions. The other would have a longer lifetime and more complex
decay modes.[9]

A rough justification of the argument goes as follows. According to equation (3.1), charge conjugation (particle–antiparticle interchange) transforms K_1^0 into itself. Similarly, a spin-zero system of two pions transforms into itself. Thus, if charge conjugation is conserved in the weak interactions, only K_1^0 can decay into two pions. K_2^0, which goes into $-K_2^0$ under charge conjugation, cannot decay into two pions.

The switch from K^0, $\overline{K}^0$ to K_1^0, K_2^0 was more than a change in representation. Physicists require that elementary particles have well-defined masses and lifetimes: Because K_1^0, K_2^0 enjoy these properties, they (not K^0, $\overline{K}^0$) should be considered the particles. Moreover, neutral K mesons are usually observed by their decays; so, again, K_1^0, K_2^0 have preference. Nevertheless, in a strong interaction such as $\pi^- p \rightarrow K^0 \Lambda^0$, where strangeness is conserved, physicists will speak of production of a K^0 meson, although to explain subsequent observation they refer to superposition of K_1^0 and K_2^0. In particular, the K^0 particles thus produced give a beam of oscillating strangeness, because the K^0 can transform into a $\overline{K}^0$: The K^0 does not retain all of its properties, in particular its strangeness, violating a usual requirement for a particle. The physics community has accepted the odd features of neutral K-meson behavior with equanimity and without much comment.

The earliest reports of neutral kaons concerned the two-pion decay attributed to K_1^0. There had been earlier reports[10] of "anomalous" V^0 events that could not be assigned to this decay, further complicating the picture. With the work of Gell-Mann and Pais, however, it became reasonable to attribute at least some of these events to their long-lived K_2^0. In 1956, Lande and associates detected the predicted long-lived K mesons.[11] They exposed a magnetic cloud chamber to a neutral beam six meters from the target, which represented 100 mean lives for the Λ^0 and K^0 particles in the beam. They observed twenty-six events that could not be interpreted as $K^0 \rightarrow \pi^+ \pi^-$, and they decided that although the events were consistent with decay of a K^0 into three particles, their results did not establish identity between the observed particles and the K_2^0 meson predicted by Gell-Mann and Pais.

Other experiments were also consistent with the particle-mixture theory.[12] For example, Gell-Mann and Pais had also predicted that one-half of all K^0 decays should be via K_2^0. In 1957,

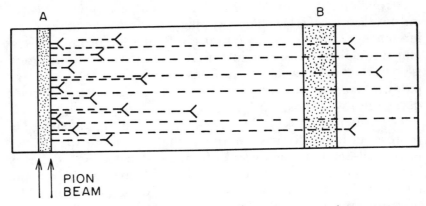

Figure 3.1. Schematic diagram of the experiment to show K_1^0 regeneration suggested by Pais and Piccioni.[14]

F. Eisler and associates found that 51 ± 7.5 percent of the K^0 mesons they observed did not decay to K_1^0, and they identified these decays "with the K_2^0 meson proposed by Gell-Mann and Pais." Additional evidence similarly evaluated was provided by the group's finding of an event that fit the reaction $\pi^- p \rightarrow \Sigma^0 K^0$. The K^0 turned up at a distance 10 K_1^0 lifetimes from the place of production. Also, it was accompanied by two charged decay particles on one side of its direction of flight, indicating that it could not decay as a K_1^0.[13] Nevertheless, this evidence, like Lande's, was consistent with the predictions of Gell-Mann and Pais, without testing the particle-mixture theory.

One striking effect that would have established the theory was regeneration of K_1^0 mesons, as predicted by Pais and Piccioni.[14] They argued that the strong interactions of K^0, $\overline{K}^0$ would be quite different. Because the reaction $\pi^- p \rightarrow K^0 \Lambda^0$ has a sizable cross-section, the processes $\overline{K}^0 p \rightarrow \Lambda^0 \pi^0$ and $\overline{K}^0 n \rightarrow \Lambda^0 \pi^0$ must also be quite probable. Strangeness conservation in the strong interactions, however, forbids the equivalent reactions for K^0, because K^0 and Λ^0 have opposite strangeness. In passing through matter, the $\overline{K}^0$ should be absorbed more strongly than the K^0.

To confirm that prediction, they proposed the experiment illustrated in Figure 3.1. A beam of pions incident on target A produces K^0 mesons at 90°. The K^0 can be represented as $|K^0\rangle = (|K_1^0\rangle + |K_2^0\rangle)(1/\sqrt{2})$. After a time that is long relative to the K_1^0 lifetime, the beam at 90° will consist almost entirely of K_2^0 mesons, because the K_2^0 lives much longer than the K_1^0. We can then represent the beam as $|K_2^0\rangle(1/\sqrt{2} = (|K^0\rangle - |\overline{K}^0\rangle)/2$. When the K_2^0 interact with the thick absorber B, most of the $\overline{K}^0$ component is removed, while the K^0 component remains unchanged. The emergent beam has a restored K_1^0 component one-fourth the intensity (one-half the amplitude) of the original K^0 beam: $|K^0\rangle/2 = (|K_1^0\rangle + |K_2^0\rangle)/(2\sqrt{2})$. Therefore, two-pion decays should reappear, with about one-fourth the population observed near the target A. The predicted regeneration of K_1^0 was confirmed by a large group working with Muller in 1960. Furthermore, they confirmed M. L. Good's deduction that the regeneration would be coherent.[15]

A second consequence of the particle-mixture hypothesis of Gell–Mann and Pais is the possibility of observing K_1^0–K_2^0 interference between K_1^0 and K_2^0. Assuming that the mesons differ in lifetimes and in masses, we can write the probability of finding a K^0 at time τ as

$$P(K^0, \tau) = \frac{1}{4}[e^{-\gamma_1\tau} + e^{-\gamma_2\tau} + 2e^{-(\gamma_1+\gamma_2)\tau/2}\cos(m_2 - m_1)\tau]$$

$$P(\overline{K}^0, \tau) = \frac{1}{4}[e^{-\gamma_1\tau} + e^{-\gamma_2\tau} - 2e^{-(\gamma_1+\gamma_2)\tau/2}\cos(m_2 - m_1)\tau]$$

where γ refers to the inverse lifetimes of the mesons, and m refers to their masses.

Two methods were proposed to detect this interference. The earliest, suggested by Zel'dovich and by Treiman and Sachs, involved looking at the semileptonic decays of K^0: $K^0 \rightarrow \pi^\pm e^\mp \nu$ and $K^0 \rightarrow \pi^\pm \mu^\mp \nu$.[16] According to the rule $\Delta S = \Delta Q$, $\pi^+ e^- \nu$ and $\pi^+ \mu^- \nu$ decays can result only from $\overline{K}^0$, and $\pi^- e^+ \nu$ and $\pi^- \mu^+ \nu$ can result only from K^0 decays. Here physicists refer to K^0 and $\overline{K}^0$ rather than to K_1^0 and K_2^0. For an initially pure K^0 beam, the expected time variation for these decays (and thus for the $\overline{K}^0$, K^0 intensities) is shown in Figure 3.2. The frequency of the "beat oscillations" or interference term depends on Δm. The second method, by Fry and Sachs,[17] required detecting the strangeness oscillation of a kaon beam by looking at strong interactions that could be produced only by $\overline{K}^0$; for example, $\overline{K}^0 p \rightarrow \Lambda \pi^+$.

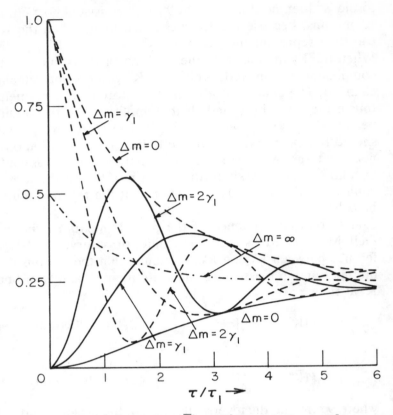

Figure 3.2. Probability of finding a $\overline{K}^0$ or a K^0 in an initially pure K^0 beam as a function of time, for several values of the K_1^0–K_2^0 mass difference Δm: dash line, $P(K^0)$; solid line, $P(\overline{K}^0)$. From Kabir.[2]

In 1957, the situation changed dramatically. In order to resolve the θ–τ puzzle, Lee and Yang had suggested that parity and charge-conjugation invariance were violated in the weak inter-actions. They offered several possible experimental tests, and three experiments soon confirmed both suggestions (see Chapter 1 for details). That confirmation would seem to have undermined the particle-mixture theory of K^0 mesons, which assumed invar-iance under charge conjugation. That was not the case, however:

The existence of degenerate but distinct K^0 and $\overline{K}^0$ states does not rest on C-invariance but is a consequence of the strangeness possessed by the K^0 and a much more general symmetry whose existence was stated in the form of a theorem by Lüders and Zumino. This remarkable theorem . . . states that any quantum field theory which incorporates the conditions of relativistic invariance and locality is automatically invariant under an operation which can be identified with the combined operation TCP of space-inversion, charge-conjugation, and time-reversal (more accurately, motion reversal), whether or not the theory is invariant under any of these transformations separately. A particular consequence of this theorem which we have mentioned already, is that every particle must have an antiparticle (which may coincide with itself if conservation laws permit) with the same mass and lifetime and with all "charges" reversed.[18]

In 1957, Lee, Oehme, and Yang further examined the consequences of noninvariance under time reversal and charge conjugation.[19] They supposed that the TCP theorem and also the operations T, C, and P held separately for the strong interactions, but not necessarily for the weak interactions. They pointed out that under these assumptions, K_1^0 and K_2^0 could decay into the same final states, and thus K_1^0–K_2^0 interference could be observed by looking at these decays. Most of the features of the theory of Gell-Mann and Pais remained. Another consequence was that if one of the three operators P, C, T failed of conservation, at least one other must also fail. Should CP fail, the ratio $(K_2^0 \rightarrow e^- \pi^+ \nu)/(K_2^0 \rightarrow e^+ \pi^- \nu)$ would not necessarily be unity. This would also be true for the decays $K_2^0 \rightarrow \mu^\mp \pi^\pm \nu$.

Even before the experimental confirmation of parity violation, Landau had pointed out some of the happy consequences of looking at the combined operation CP:

At first glance it appears that non-conservation of parity would imply an asymmetry of space with respect to inversion. Such an asymmetry, in view of the complete isotropy of space (conservation of momentum) would be more than strange. In my opinion a simple denial of parity-conservation would place theoretical physics in an unhappy situation. I wish to point out that there exists a way out of this situation. We know that the strong interactions are invariant not only with respect to space-inversion but also with respect to charge-conjugation. We assume that in the weak interactions these two invariance properties do not hold separately. But we can suppose that we still have invariance with respect to the product of the two operations [CP], which we call combined inversion. Combined inversion consists of space reflection with interchange of particles and antiparticles. If all interactions are invariant with respect to combined inversion, space remains completely symmetrical and only electric charges are asymmetrical. This asymmetry destroys the symmetry of space just as little as the existence of chemical stereoisomers.[20]

Landau noted that the theory of Gell-Mann and Pais would still apply, although with CP rather than C invariance. Conserved parity is the product of ordinary parity and charge parity. K_1^0, which makes two pions, is thus an even particle, and K_2^0 is an odd particle. A further interesting consequence was pointed out by Treiman and Wyld:[21] If the TCP theorem is true, then should CP fail, so must time reversal.

In 1957, most of the work in the field of weak interactions, both experimental and theoretical, was devoted to examining the implications of parity nonconservation and the $V-A$ theory of weak interactions, which incorporated it in a natural way. Nevertheless, some work was done on CP conservation. In 1958, M. Bardon and associates pointed out that "observation of a long-lived neutral K meson for which 2 pion decay is forbidden would be evidence for time reversal invariance in the decay interaction," and they set an experimental upper limit of 0.6 percent on $K_2^0 \rightarrow \pi^+ \pi^-$. They made it clear, however, that no single result had yet "demonstrated the necessity of the K^0 particle mixture, although it is at present the only theory sufficient to explain all the observations."[22]

Several theorists concentrated on the formal implications of CP invariance. J. J. Sakurai suggested that one might look for polarization of the muon in $K_{\mu 3}$ decay as a possible test of time reversal and thus of CP invariance.[23] Weinberg pointed to experiments showing that the K_2^0 decays much more slowly than the K_1^0 and that the K_2^0 seldom if ever decays to two pions: "This is just what one would expect if CP (or C) were conserved in K^0 decay. Conversely we may ask how much support is given to CP invariance by these experiments and what additional support may be gained by similar experiments in the near future."[24] He suggested looking at charge asymmetries in the processes $K_2^0 \rightarrow \pi^\pm e^\mp \nu$ and $K_2^0 \rightarrow \pi^\pm \mu^\mp \nu$ as a further test, although he expected the asymmetries to be small. M. L. Good observed that the absence of $K_2^0 \rightarrow 2\pi$ shows that the gravitational masses of K^0 and $\overline{K}^0$ cannot differ by more than a few parts in 10^{-10} of the K^0 inertial mass and that the gravitational mass cannot depend on the strangeness quantum number, which has different values for K^0 and $\overline{K}^0$.[25]

In 1960, the discussion entered a new phase, when Muller's group observed regeneration of K_1^0 mesons and corroborated the particle-mixture theory. D. Neagu and associates then further tested CP invariance by examining the charge ratio of leptonic decays; they found 46 π^- and 51 π^+ events, a ratio of 0.90 $\pm$ 0.18. No examples of $K_2^0 \rightarrow \pi^+ \pi^-$ turned up: "Our results on the charge ratio and the degree of 2π-decay forbiddeness . . . provide no indications that time reversal invariance fails in K^0 decay."[26] Sachs agreed, but observed that neither did the absence of $K_2^0 \rightarrow 2\pi$ provide clear-cut proof of the validity of CP.[27] He did not, however, discuss the implication that should $K_2^0 \rightarrow 2\pi$ be observed, CP would be violated. Further experiments pushed the upper limit for a two-pion decay of K_2^0 to less than 0.3 percent of K_2^0 decays. No evidence of charge asymmetry in the decays $K_2^0 \rightarrow e^\pm \pi^\mp \nu$ presented itself.[28]

The immediate background for the Princeton experiment was the report by Adair and associates of anomalous regeneration of K_1^0 mesons from K_2^0 mesons. They guessed that the regeneration resulted from a new weak long-range interaction, from an "anomalous coherent production." They emphasized the "extraordinary consequences which may be required by such a result" and suggested that perhaps it was only an artifact of a "combination of real effects underestimated by us together with strong statistical fluctuations."[29] It was an artifact: A spurious result stimulated the work of the Princeton group.

3.2 THE PRINCETON EXPERIMENT

Cronin, Fitch, and Turlay outlined the purposes of their experiment in their proposal to the Brookhaven National Laboratory:

The present proposal was largely stimulated by the recent anomalous results of Adair, et al., on the coherent regeneration of K_1^0 mesons. It is the purpose of this experiment to check these results with a precision far transcending that attained in the previous experiment. Other results to be obtained will be a new and much better limit for the partial rate of $K_2^0 \rightarrow \pi^+ \pi^-$, a new limit for the presence (or absence) of neutral currents as observed through $K_2^0 \rightarrow \mu^+ \mu^-$.[30]

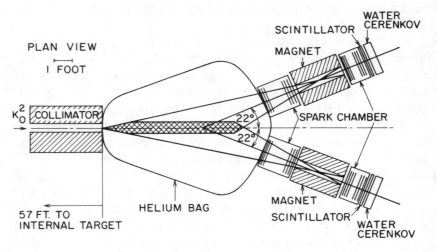

Figure 3.3. Plan view of the apparatus used by the Princeton group to detect $K_2^0 \rightarrow \pi^+\pi^-$. From Christenson et al.[1]

These experimenters recognized the relevance of their experiment for the question of $K_2^0 \rightarrow \pi^+\pi^-$ decay (and thus for CP conservation). They did not, however, expect to observe that decay. A preliminary estimate indicated that the CP phase of the experiment would detect around 7,500 K_2^0 decays, and thus the limit on CP violation could be reduced from 1/300 to 1/7,500 if no two-pion decays were detected. "Not many of our colleagues would have given credit for studying CP invariance but we did so anyway."[31]

The relative importance of the phase of the experiment testing CP invariance is further indicated by the fact that of 140,000 pictures taken during the experiment, about 50,000 obtained with no regenerator in the beam could be used to search for $K_2^0 \rightarrow \pi^+\pi^-$, the CP-violating decay.[32] The remaining pictures had either a dense regenerator to study coherent regeneration and the K_1^0–K_2^0 mass difference or a liquid-hydrogen regenerator to check the results of Adair. The testing of CP invariance was a deliberately planned and important part of the experiment.

The experimental apparatus is shown in Figure 3.3. The products of a beam of K_2^0 mesons, decaying in a volume filled with helium, were detected in two spectrometers, each consisting of spark chambers, scintillation counters, magnet, and a water Cerenkov counter. The scintillation counters guaranteed that a charged particle had passed through each spectrometer. The water Cerenkov counter ensured that the particle was an electron, muon, or pion. The spark chambers were triggered by a coincidence between the counter telescopes on each side to make sure that the particles derived from a single K_2^0 decay. The spark chambers delineated the particle tracks in the magnetic fields, from which the vector momentum of each decay product could be calculated.

The notebook for the experiment contains a record of equipment malfunctions, human error, unexpected difficulties, and interference of the experiment with other experiments being run concurrently at the accelerator. It includes a record of the checks performed on the proper functioning of the apparatus, which the physics community conventionally assumes to have been done, but which usually are not discussed extensively in published papers. The book covers the period 2 June 1963 to 22 July 1963, during which all data were taken. Before 2 June, the experimental apparatus, including magnets, collimators, spark chambers, Cerenkov counters, and scintillation counters, had been surveyed into place, and operating conditions for the counters and spark chambers had been determined.

The early phase of the experiment consisted in setting up beam monitors, particularly one to measure neutron intensity as a measure of K^0 intensity. That monitor went through four different configurations as successive experimental difficulties obtruded. An entry for 6 June reads: "The above is absolute garbage." On the same day, several modifications later: "Finally think we have a neutron monitor." The monitor was checked against an already calibrated device situated in an experiment located on the same beam line as the Princeton experiment but farther from the accelerator. "Ratio of neutron monitor to that of Jovanovic seems quite constant (to 3%) ← statistics." This monitor, labeled Jovanovic for one of the physicists in the other group, provided a constant check on the operation of the Princeton neutron monitor.

Interaction with other experiments was not always so beneficial. Again, on 6 June: "These runs were interrupted by discovery that bending magnet of Frisch at 6 BeV gives ~20/1 ratio of [counter] 3 to [counter] 2. This is intolerable. Now they have reduced beam

and we resume running, pending solution." The setting of a magnet in the adjacent experiment run by an MIT group drastically changed the operation of the Princeton experiment. The
need for, and occurrence of, cooperation is indicated by the
reduction in the MIT beam intensity. On 10 June, and again on
14 June, the experimenters installed lead shielding to cut out
the Frisch beam, and they constantly recorded the setting of the
MIT magnet in case problems occurred during analysis of the
data. On 14 June, the welcome news that "MIT gone back to
MIT" appeared. Later in the run, on 10 July, the Jovanovic
group installed absorbers downstream of the Princeton experiment. Although the move did not seem to have noticeable effects, the changing conditions of the Jovanovic absorber were
also carefully recorded.

This careful checking on the uniform operation of the experiment is indicated in a typical data page (Figure 3.4). It gives the
time a run began, the event numbers at the beginning and end
of the run, the counts in the two Cerenkov counters, the readings
on the neutron and Jovanovic monitors, the numbers of beam
pulses and picture-taking rates per unit neutron intensity, and the
ratio of the readings of the neutron and Jovanovic monitors. The
constancy of these ratios served as a check on the uniform operation of the apparatus. A final column records experimental
conditions, such as regenerator used, as well as comments concerning the operation of the experiment.

These comments indicate differences between the experiment
as done and as published. They refer to problems with, or loss
of, the accelerator beam, power failures (22 June), and adverse
weather (26 June): "MG set overheating because of hot day (90°–
95°)." The weather posed an unexpected difficulty when a nearby
lightning strike caused the loss of much of the data for a run:
"Lightning struck, ending this run prematurely. All scalers went
off losing their information" (28 June). Although this was a rare
occurrence, the group remembered it and noted, "Weather looks
bad – write down data often – but record in book every 50 Neu-
trons" (20 July).

Equipment failures, very seldom mentioned in any published
account of an experiment, occurred frequently. Sometimes the
experimenter's frustration with a recalcitrant piece of apparatus

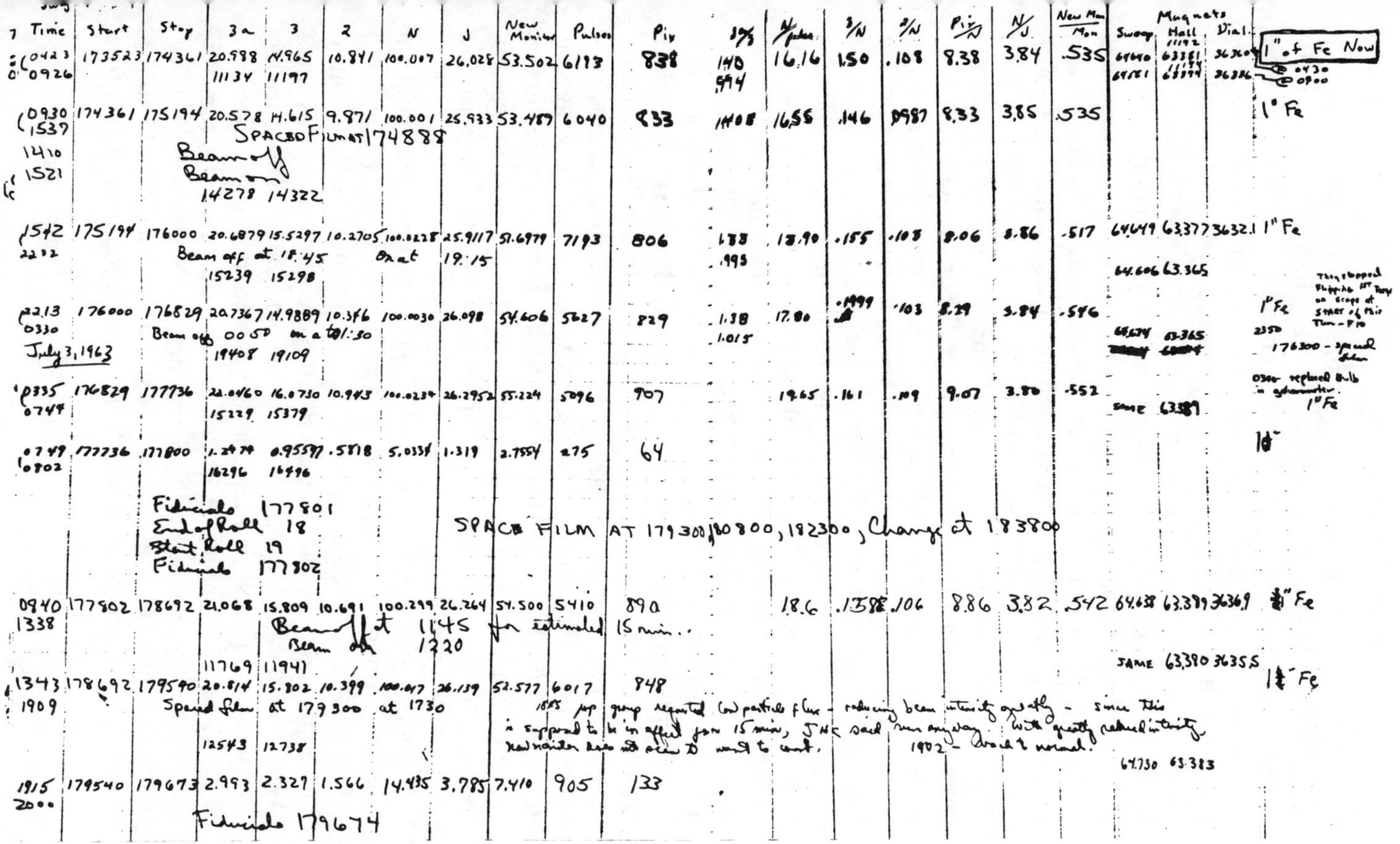

Figure 3.4. A typical page from the laboratory notebook of the Princeton group.

shows: "Shut off at 1519 . . . to fix damn fiducials" (23 June). Sometimes malfunctions gave rise to in-group humor. One of the strip mirrors used to view the spark chambers fell off. Because it blocked only one gap and could not be repaired easily, nothing was done about it. The problem got worse: "When I did check for the mirrors I found 2 mirrors fallen in the view of no. 3. One in the large S. C. [spark chamber], one in the small one [a subsequent note on the same day changed this to two in the small chamber]. We have to be more serious about these checks because it seems that the rate of falling mirrors increases!!!" (20 July). Later that day, another wag wrote: "Only 3 strip mirrors have fallen to date. . . . The SMFR [strip mirror falling rate] has not increased so alarmingly – there is hope." Sometimes there is no apparent reason for a malfunction. "Neutron monitor had just shut itself off" (27 June). "Stopped to fix N [neutron monitor] or J [Jovanovic monitor] – Trailer people [another group] have a perfect ratio of N/J – so, trouble must be in our 2nd discriminators or scalers. Checked discs [discriminators] – nothing obviously wrong. Start again – seems okay" (10 July). The malfunction had cured itself. The malevolence of the inanimate: "During last run it was noticed that N and J did not agree. . . . Perhaps a scalar was not properly reset etc. In any case it is difficult to say which monitor was off. The probability is large that it is J" (13 June). Later that day: "Found neutron scaler turned off (gate closed) – chamber must have done it. So merely turned it on. This suggests this is what happened in run above. Now we have two scalers on neutron monitor – count same within .1%." Note the careful checking of the operation of the apparatus and the safeguard put in to avoid a problem.

Human errors also occurred: "H. P. knocked one of the neutron counters on floor and broke scint [scintillator] to light pipe joint – I merely reinstalled counter to finish out morning" (9 June). The equanimity of this comment is typical but not universal: "While checking logic flip-flops, found some damned boat-rocker had interchanged Nixie lites and Reg. Adv. outputs – Lites are 25 ms, Reg is 55 msec – Luckily, 25 ms will drive the register, but just barely. So, since this happened all Nixies will be overexposed – can even find culprit by exploiting this fact" (18 June). There is an occasional apology: "Film ran out sometime (?) before this – noticed it at this point – somewhat before 131700 it would seem! Sorry" (20 June). The complex equipment sometimes presented spurious malfunctions: "Noticed system #2 camera appeared not to be advancing film – stopped run – entered 2 long pulses – all ok – probably all was ok anyway. Film was advancing previous panic uncalled for – perfect symmetry of take up reel made it difficult to tell when film was advancing and I jumped to conclusions – so except for 2 unnecessary long pulses – nothing lost" (26 June).

The test of CP was carefully performed and analyzed using the experimental apparatus pictured in Figure 3.3. The vector momentum of each of the two charged decay products from the K_2^0 beam and the invariant mass m^* were computed, assuming each product had the mass of a pion: $m^{*2} = (E_{\pi 1} + E_{\pi 2})^2 - (\mathbf{P}_{\pi 1} + \mathbf{P}_{\pi 2})^2$. If both particles were indeed pions from K_2^0 decay, m^* would equal the K_2^0 mass. They also computed the vector sum of the two momenta and the angle θ between this sum and the direction of the K_2^0 beam. This angle should be zero for two-body decays, but not in general for three-body decays.

The experimental distribution in m^* for the K_2^0 decays in helium appears in Figure 3.5(a) along with the results of a Monte Carlo calculation. For events with m^* between 490 and 510 MeV, Figure 3.5(b) shows the distribution in $\cos \theta$ along with the corresponding Monte Carlo results. A clear peak above the Monte Carlo is seen for $\cos \theta > 0.9999$ ($\cos \theta \approx 1$ means $\theta \approx 0$). Events with $\cos \theta > 0.9995$ were remeasured on a more precise machine and independently recomputed. Figure 3.6 shows the results from the more accurate remeasurement in three mass regions, one above, one below, and one encompassing the K^0 mass (498 MeV). For events with $\cos \theta > 0.99999$, the average mass fell out as 499.1 ± 0.8 MeV, in good agreement with calibration runs on the decay of K_1^0 mesons coherently regenerated in tungsten. Similarly, the angular deviation, with a forward peak with a standard deviation of 4.0 ± 0.7 milliradians, agreed with the tungsten data. For these events (with $\cos \theta \approx 1$) they also computed the mass of one decay particle assuming K^0 decay and assuming that the other was a pion. They found the mass to be 137.4 ± 1.8 MeV, in good agreement with the charged-pion mass, 139.6 MeV. These results in-

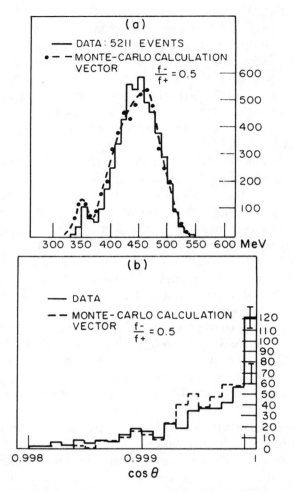

Figure 3.5. (a) Experimental distribution in m^* compared with the Monte Carlo calculation. The calculated distribution is normalized to the total number of observed events. (b) Angular distribution of events in the range $490 < m^* < 510$ MeV. The calculated curve is normalized to the number of events in the complete sample. From Christenson et al.[1]

dicated that the decay events were identical with the regenerated K_1^0 events and thus could be identified as K_2^0 decays. After background subtraction, Cronin's group obtained a total of 45 ± 9 $K_2^0 \to \pi^+ \pi^-$ events in a corrected sample of 22,700 K_2^0 decays.

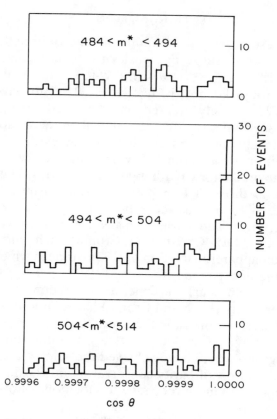

Figure 3.6. Angular distributions in three mass ranges for events with $\cos \theta > 0.9995$. From Christenson et al.[1]

They also examined other possibilities that might have led to a forward peak in the angular distribution at the K^0 mass, including K_1^0 coherent regeneration, $K_{\mu3}$ or K_{e3} decay, and $\pi^+\pi^-\gamma$ decay, and rejected them all. Any alternative, they wrote, required "highly nonphysical behavior of the three body decays of the K_2^0." They declared the reality of the decay $K_2^0 \rightarrow \pi^+\pi^-$, and violation of CP. Statistically their case was quite strong. The 45 ± 9 events is a 5-S.D. effect. The probability that it was not a statistical artifact is 1.7 million to 1.

3.3   THE IMMEDIATE REACTION, 1964–5

The Princeton results were almost immediately confirmed, albeit with lower statistics: a suggestive peak of eleven events, which suggested that $K_2^0 \to \pi^+ \pi^-$ occurs in 0.2–0.3 percent of all K_2^0 decays.[33] The group did not, in fact, even claim to have seen $K_2^0 \to \pi^+ \pi^-$. They merely pointed out that their results were consistent with such a decay. Most theoretical physicists accepted the Princeton results and their confirmation as evidence either for CP violation or for apparent CP violation, and they offered explanations and predictions of other possible oddities. The standard line, taken in three of four papers: "The existence of the decay mode $K_2^0 \to \pi^+ \pi^-$ has recently been reported by Christenson, Cronin, Fitch, and Turlay. This establishes the violation of CP invariance."[34] Or: "Christenson, Cronin, Fitch, and Turlay have reported an apparent violation of CP invariance which manifests itself in the decay mode $K_2^0 \to 2\pi$."[35] The reasons for this quick acceptance seem clear. The experiment was done by a group whose senior members (Fitch and Cronin) had a good reputation for careful and correct work, and their result, 45 ± 9 events, was statistically persuasive. (It is doubtful that the confirming results of Abashian's group would have achieved the same effect.) Perhaps also the overthrow of parity conservation in 1957 had made the physics community more receptive to violations of other discrete symmetries, like CP.

The remaining 25 percent of published opinion offered alternative explanations for the Princeton results, explanations that sought to preserve CP conservation. These alternatives, which show the reluctance of the physics community to accept refutation of an important and strongly held principle, relied on one or all of three arguments: (1) The Princeton results were caused by a CP asymmetry (the local preponderance of matter over antimatter) in the environment of the experiment. (2) $K_2^0 \to \pi^+ \pi^-$ does not necessarily imply CP violation. (3) The Princeton observations did not arise from $K_2^0 \to \pi^+ \pi^-$ decay. This last argument can be divided into the assertions (3a) that the decaying particle was not a K_2^0, (3b) that the decay products were not π mesons, and (3c) that another unobserved particle (similar to the neutrino in β decay) came off in the decay.[36]

More speculatively, we might classify these alternative explanations by the degree of commitment of the authors to their truth. Many are what might be called interesting speculations rather than serious suggestions. This is not to say that they were frivolous, but rather that they were offered in the spirit of what Popper has called "bold conjectures," without a strong commitment to their truth.[37] They show the physics community exercising its intellectual muscles to produce ingenious solutions to an important and interesting problem, which solutions might also, though improbably, preserve an important physical principle. If we learn from our mistakes in the way Popper supposes, this type of speculation is an important part of the progress of science. It does appear that periods of speculation follow odd discoveries, such as the discovery of the Ψ particle. The Ψ appeared to be produced copiously, but to decay rather slowly, just like the K meson. Various explanations came forth, including the suggestion of two new quantum numbers "charm" and "color." The solution eventually accepted resembled the solution of "strangeness" given for the K mesons.[38] Similarly, the discovery of X rays gave rise to many speculative theories.[39] Before discussing these alternatives to CP violation, it is worth examining why they might have been offered.

In the case of parity nonconservation (and, with it, violation of charge-conjugation invariance), Landau's suggestion that CP was the appropriate operation to consider tied the violation of mirror symmetry (P) to a real structure in the world, namely, the existence of particles and antiparticles. No such alternative or explanation was available for CP violation. In addition, there were strong reasons for believing in the CPT theorem; violation of CP invariance implied that of time-reversal invariance, a conclusion unappealing to most physicists:

The invariance of nature with respect to the operation of combined inversion and the related conservation of CP parity – this is the beautiful hypothesis first proposed by Landau to preserve the invariance of empty space (vacuum) under spatial inversion. It arose after it was established in 195[7] that spatial parity (P) is not conserved in weak interactions. The denial of CP conservation (and the associated breakdown of T invariance of nature) is a difficult step for theoretical physics. Our vaunted "common sense" does not permit us to picture isotropy of space-time with a preferred direction of the time axis. The connection between conservation of CP and T parity comes from the CPT theorem which has a very profound basis in theory. The principles of special relativity and the connection between spin and statistics (which is itself a consequence of the theory of relativity and the positive character of energy) automatically lead to conservation of CPT. . . . Generally speaking, it may nevertheless turn out that CPT parity is not conserved, and then the violation of CP invariance observed in the experiment of Christenson, Cronin, et al. will not mean nonconservation of time (T) parity. Only future experiments can tell us the correct answer. At the moment it is natural to want to choose the lesser of two evils. In the mathematical structure of our present relativistic physics there is literally no place for violation of CPT invariance. It is therefore highly probable that the experiment of the Princeton group also indicates the absence in nature of invariance with respect to time inversion $t \rightarrow -t$ or, as we say, violation of T invariance.[40]

At that time there was reasonably good experimental evidence supporting the CPT theorem's prediction of the equality of masses and lifetimes of particles and antiparticles. These included mass(π^+)/mass(π^-) = 1.0021 ± 0.0027; $\tau(\pi^+)$ = (2.53 ± 0.10) × 10^{-8}sec and $\tau(\pi^-)$ = (2.55 ± 0.19) × 10^{-8}sec; $m(\bar{p})/m(p)$ = 1.008 ± 0.005; $\tau(\mu^+)$ = $\tau(\mu^-)$ to an accuracy of 0.2%.[41] In addition, as Good observed, the absence of $K_2^0 \rightarrow 2\pi$ showed that the $\bar{K}^0$ and K^0 masses were equal to one part in 10^{10}.[42] There was then, and even to this day, no observation that showed time reversal directly. The best evidence for CP conservation was the absence of $K_2^0 \rightarrow 2\pi$ decay, a point emphasized by both Prentki and Terent'ev.[43] Given the theoretical and experimental support for the CPT theorem, and the lack of evidence either way for T violation or CP violation (except $K_2^0 \rightarrow 2\pi$), physicists, not unnaturally, hesitated before accepting the death of CP.

Another cause for hesitation was the complex nature of the K^0 system, which included the particle-mixture (K_1^0, K_2^0) theory of Gell-Mann and Pais, as well as the unique phenomenon of K_1^0 regeneration predicted by Pais and Piccioni and observed by Muller. Observation of $K_1^0 \rightarrow 2\pi$ decay took physics into a realm previously unexplored, into "an entirely new domain of phenomena, those involving extremely small energies."[44] The Princeton experiment could detect energy differences around 10^{-8} eV, which no other experiment in particle physics could do. Still, none of the other experiments that might have shown CP violation did so. "Therefore it is not unnatural to try to explain the new experiments in some other way."[45]

The fact that the observed effect was so small in absolute terms [the branching ratio $(K_2^0 \rightarrow \pi^+\pi^-)$/(all K_2^0 decays) = 0.002 ± 0.0004] may have encouraged both theoretical speculation and further experiments to corroborate the initial effect.[46]

The first alternative explanation, and one of the most interesting, was the cosmological model proposed independently by Bell and Perring and by Bernstein, Cabibbo, and Lee.[47] They proposed a long-range interaction of cosmological origin, a hyperphoton, that interacted differently with K^0 and $\overline{K}^0$ and thus could convert K_2^0 mesons into K_1^0 mesons in a way similar to the regeneration described earlier. They estimated that the effect of this vector field would be quite small, on the order of 10^{-17} of the K^0 mass, something quite undetectable by experiments of the Eötvös type. An important consequence of the cosmological model was that the rate of $K_2^0 \rightarrow 2\pi$ should be proportional to the square of the energy of the decaying K_2^0.

These suggestions belong to the class of interesting speculations. "Before a more mundane explanation is found," Bell and Perring wrote, "it is amusing to speculate that it might be a local effect due to the dys-symmetry of the environment, namely the local preponderance of the matter over antimatter."[48] And both Lee and Cabibbo proposed alternative solutions to their alternative solution and tried to *explain* the origin of CP violation.[49] And Lee proposed still another model to *avoid* CP violation.

Even speculations are open to criticism, and Weinberg[50] immediately pointed out a serious difficulty with the cosmological model. Because neither strangeness nor isotopic spin, the supposed origins of the field, is absolutely conserved, then the hyperphoton must have a finite mass, related to the range of the interaction. Assuming that the range of the interaction was the size of our galaxy, he calculated the ratio $(K_2^0 \rightarrow 2\pi + $ hyperphoton$)/K_2^0 \rightarrow 2\pi)$ as 10^{19}. This implied that the K meson and all strange particles would be totally unstable, which Weinberg regarded as a catastrophe. Other theoretical objections were raised by Lyuboshitz, Okonov, and Podgoretskii,[51] who argued that the models of Bell and Lee "contradict the principle of 'field interaction' on which our present physical concepts are based" and that, if the theory were made consistent with this principle, it would contradict results of Eötvös experiments. Because of the smallness of the effect, however, these models could not be tested in such experiments.[52]

Lee's second interesting speculation required an interaction that involved the gradient of a scalar field.[53] A consequence of the theory was that the rate of $K_2^0 \rightarrow 2\pi$ should vary as the square of the K^0 energy. Marx pointed out that the energy density of the universe required by this model would be eight orders of magnitude larger than that allowed by relativistic cosmology.[54]

The theoretical objections to these models became irrelevant when two experiments measured the $K_2^0 \rightarrow 2\pi$ rate at energies different from each other and from that of the original Princeton experiment and failed to find the predicted energy dependence. The first, by De Bouard,[55] measured the $K_2^0 \rightarrow 2\pi$ rate at a mean momentum of 10.7 GeV/c and found it to be $(3.5 \pm 1.4) \times 10^{-3}$ [changed to $(2.24 \pm 0.23) \times 10^{-3}$ in a later report] of all charged decay modes. Their result conflicted with the value of $(1.6 \pm 0.3) \times 10^{-1}$ expected using the branching ratio measured by the Princeton group at a momentum of 1.1 GeV/c and the assumed energy dependence. The second, by Galbraith,[56] at a mean momentum of 3.1 GeV/c, obtained $(2.08 \pm 0.35) \times 10^{-3}$, compared with an expected value of $(13.4 \pm 3.0) \times 10^{-3}$. These experiments clearly ruled out a rate that depended on the square of the K^0 energy. In addition, these measurements related to decays in a vacuum, where regeneration effects need not be feared, and both sets of experiments claimed to have established that the decay particles were pions.

Other models, proposed by Grib[57] and by Marx,[58] avoided making the decay rate dependent on the square of the energy and attributed the observed decay to the fact that "vacuum," the locus of the decays, is not an eigenstate of CP. These models did not predict the existence of any other "apparent" CP violations and hence could not be tested. In addition, there does not seem to be any major distinction in our fundamental ideas about symmetry whether we attribute the apparent violation of CP invariance to the weak interactions or to the universe or physical vacuum. These models, therefore, presuppose the fact of CP violation.

Three other models also used an external field to explain the apparent CP violation. Kundu[59] and Spitzer[60] both discussed the possibility of K_1^0 regeneration from K_2^0 by an external magnetic field. Kundu noted that the rate for this process would also vary as the square of the energy, thus conflicting with the experimental results discussed earlier. A similar objection held for the model of Kabir and Lewis, in which regeneration occurs through the interaction with "invisible" particles, for example, a neutrino sea.[61]

Another type of alternative explanation questioned that the process observed was $K_2^0 \rightarrow \pi^+ \pi^-$ decay. One model of this type, proposed by Levy and Nauenberg,[62] supposed for the process $K_2^0 \rightarrow K_1^0 + S$, where S is a neutral particle of mass less than the K_1^0–K_2^0 mass difference, followed by $K_1^0 \rightarrow 2\pi$. Because the K_1^0 lives a much shorter life than the K_2^0, and because the S particle would not be observed, experiment would show an apparent $K_2^0 \rightarrow 2\pi$ decay. Similarly, Kalitzin[63] suggested $K_2^0 \rightarrow \pi^+ \pi^- + S$, where S, the "paritino," is analogous to the neutrino of β decay. Both models fell when Fitch and his collaborators[64] observed interference between $K_2^0 \rightarrow 2\pi$ and $K_1^0 \rightarrow 2\pi$ when the K_1^0's were produced by coherent regeneration from K_2^0's. Although initially in phase, the two wave functions K_1^0 and K_2^0 have different time dependences because of the difference in masses of the particles, and interference between them can be observed. Interference would not occur if another particle were emitted.

Cvijanovich, Jeannet, and Sudarshan[65] suggested that one of the pions supposedly emitted in $K_2^0 \rightarrow \pi^+ \pi^-$ decay is not a pion but rather a particle identical with it in mass but having spin 1, rather than 0, the so-called spion. This odd explanation was originally offered to explain a small asymmetry observed in $\pi \rightarrow \mu$ decay, particularly in low-energy pions emitted in $K \rightarrow 3\pi$ decays.[66] The spion would decay by both $\pi \rightarrow \mu$ and $\pi \rightarrow e$, with roughly equal rates, as opposed to a ratio of 10^4 for normal pion decays. If a neutral spion existed, it would have a preferred decay mode of $e^+ e^- \gamma$, with a very short lifetime. Rinaudo[67] looked for such effects and found no evidence for either effect. Taylor[68] also examined $K^+ \rightarrow 3\pi$ decays and found not only no asymmetry in the $\pi \rightarrow \mu$ decay but also no evidence of $\pi \rightarrow e$

decays. In addition, the spion model did not allow the interference observed by Fitch.

That observed interference also ruled out the class of explanations that attributed the observed decay to another particle, L, produced coherently with the K^0.[69] The mass of the L and the K^0 mass must be equal to 1 part in 10^{15} to explain the equality of the K_1^0–K_2^0 mass difference obtained from interference experiments and those obtained from other experiments that did not involve $K_2^0 \rightarrow 2\pi$ decay, a rather unlikely result. Also, with this model, the observed decay rate and the interference should depend both on the method of production of the K^0 mesons and on their energy. No such dependence was observed.

The model of Nishijima and Saffouri[70] explained $K_2^0 \rightarrow 2\pi$ decay by the existence of a "shadow" universe in touch with our "real" universe through the weak interactions. This model was ruled out by the interference experiments. It nonetheless came in for consideration. Everett[71] noted that if the $K^{0\prime}$ postulated by Nishijima and Saffouri existed, then a neutral shadow pion, or $\pi^{0\prime}$, should also exist, and the decays $K^+ \rightarrow \pi^+\pi^0$ and $K^+ \rightarrow \pi^+\pi^{0\prime}$ should occur with similar rates. Although the $\pi^{0\prime}$ could not interact with any apparatus in our universe, its presence could be detected by measuring the $K^+ \rightarrow \pi^+\pi^0$ branching ratio in two different experiments, one in which the π^0 was detected and one in which it was not. If the $\pi^{0\prime}$ existed, the two measurements would differ. Everett examined the evidence available at the time and could draw no firm conclusion. Callahan and Cline[72] remeasured the $K_{\pi2}$ branching ratio, detecting only the π^+, and found it to be 21.0 ± 0.56 percent, in good agreement with the results obtained detecting the π^0 of 20.7 ± 0.6 percent.[73] This ruled out the existence of the $\pi^{0\prime}$ and, by implication, the $K^{0\prime}$.

Three further proposed explanations cast doubt on some fundamental assumptions of modern physics and must be considered serious suggestions. It had already been noted that the observed rate for $K_2^0 \rightarrow \pi^+\pi^-$, measured at 300 K_1^0 lifetimes, was that expected for K_1^0 decay at 6 K_1^0 lifetimes. If the exponential-decay law failed for K_1^0 mesons, the apparent CP violation could be explained as K_1^0 decays.[74] This suggestion was tested, at least indirectly, by Fitch and his collaborators,[75] who measured the K^+ lifetime in the region 1.9–7.8 mean lives and found no violation of the exponential law.

A second suggestion, by Messiah and Greenberg,[76] was that the pions involved in the K_2^0 decay were not bosons. Although similar to Cvijanovich's proposal that the decay pions were not real pions, it was more fundamental. An axiom of modern quantum physics, well supported by experiment, is that all wave functions are either symmetric or antisymmetric under the interchange of particles: symmetric for particles with integral spin (bosons) and antisymmetric for those with half-integral spin (fermions). If the pions in K_2^0 decay were not bosons, then the decay did not necessarily indicate CP violation. Messiah and Greenberg noted, however, that $K_2^0 \rightarrow 2\pi^0$ (also CP-violating) was rigorously forbidden whether or not pions were bosons. Thus, the observation in 1967 of $K_2^0 \rightarrow 2\pi^0$ decay by Gaillard[77] and by Cronin[78] ruled out the nonboson pion. Messiah and Greenberg cannot be classified as projectors of interesting speculations in response to odd discoveries. They began a general consideration of the symmetrization postulate (that all particles are either bosons and fermions) before the results of the Princeton experiment were published. They discussed $K_2^0 \rightarrow \pi^+ \pi^-$ in a note added in proof.

A suggestion that violated yet another fundamental tenet of quantum physics, the principle of superposition, was made by Laurent and Roos.[79] Their intervention related to a continuing study of possible tests of the superposition principle. Evidence of the seriousness with which they regarded their effort to save CP by abandoning superposition is that Roos held out after virtually all the physics community had accepted CP violation. Laurent and Roos showed that (in their words and emphasis) *"a CP invariant but non-linear theory can give results in agreement with the experiments."* One consequence of their model was that K_1^0–K_2^0 interference without regeneration should not be observed. Their model fell in 1967 when Rubbia observed the prohibited phenomenon.[80]

In mid-1965, a series of review papers and review talks at conferences appeared whose authors accepted the apparent fact of CP violation and discussed various theoretical explanations, implications, and predictions of other CP-violating effects.[81] Others discussed the phenomenological description of K^0 decays and the status of experimental work in the field.[82] The most comprehensive statement on the status of alternative explanations was Prentki's talk[83] at the Oxford Conference on Elementary Particles in

September 1965. He said that the most direct interpretation of the Princeton result was CP violation in weak interactions, and he set aside many of the alternative explanations: the cosmological model and the K^0–$K^{0\prime}$ story, because of incorrect dependence of the decay rate on energy; the Levy–Nauenberg model of an extra particle in K_2^0 decay and schemes for modifying the decay pions, because of the interference results of Fitch; giving up Bose statistics, because "the price one has to pay in order to save CP becomes extremely high"; abandoning superposition or introducing a shadow universe, because they were "even more unpleasant." His conclusion: "It seems therefore reasonable to admit that CP or T conservation is not an absolute law of nature."

During 1965, other experimental results were reported confirming CP violation or time-reversal violation in K-meson semileptonic decays. Camerini[84] found results consistent with time reversal invariance in $K^+ \rightarrow \mu^+ \pi^0 \nu$. Aubert[85] and Baldo-Ceolin[86] examined $K_2^0 \rightarrow e^\pm \pi^\pm \nu$ decays and obtained results consistent with maximal CP violation, although in the latter case CP invariance was ruled out with only a 75 percent probability. Franzini[87] also looked at this decay and obtained results consistent with CP conservation and not inconsistent with CP violation.

By the end of 1965, CP violation was accepted by almost the entire physics community – M. Roos always excepted.[88] Although not all the alternative explanations of the Princeton results had been ruled out by experimental observations, those that remained were considered either implausible or more problematic than CP violation itself.

3.4   DISCUSSION

The Princeton experiment seems to be an exemplar of what one might call a "convincing" experiment. Despite the fact that acceptance of CP violation meant that physicists would have to give up either the CPT theorem or time-reversal invariance, neither of which they were eager to discard, the large majority of those working in the specialty of weak interactions accepted the Princeton result. Physicists working outside that specialty accepted the result and its consequence even more quickly than the specialists. Also, the fact that CP violation was accepted in review papers after about one year, even though all of the alternative explanations had not been explored, argues a rapidly reached consensus. Plausible reasons, both physical and methodological, for the suggestion of these alternatives or speculations have been given.

The many follow-up experiments on small effects were natural sequels to the Princeton work and argue some belief in its validity. The smallness of the effect invited experiments to confirm the Princeton result at different energies, although the relevance of those experiments to the cosmological models was known and stated. Once $K_2^0 \to \pi^+ \pi^-$ had been observed, it was natural to look for interference of K_2^0 with regenerated K_1^0, as well as for $K_2^0 \to 2\pi^0$ decays. The extremely difficult experiments on K_1^0–K_2^0 interference without regeneration and on the charge asymmetry in $K_2^0 \to e^\pm \pi^\mp \nu$ and $K_2^0 \to \mu^\pm \pi^\mp \nu$ implied a belief in CP violation. The effects expected were so small that without premising such a belief it would seem difficult to justify the experiments.

Also worthy of emphasis is the significant role that experiment played in the corroboration and refutation of hypotheses, and the close connection of theory and experiment. As Fitch said in his Nobel talk, "It is difficult to give a better example of the mutually complementary roles of theory and experiment than in telling the story of the neutral K-mesons."[89] We see the relation clearly from the discovery of the K mesons through the suggestion of conservation of strangeness by Gell-Mann and Nishijima, the particle-mixture theory of Gell-Mann and Pais, the prediction of regeneration by Pais and Piccioni, experimental corroboration through the discovery of $K_2^0 \to \pi^+ \pi^-$, and the subsequent alternative explanations and their eventual refutation by experiment.

This finding of the fruitful and complementary roles of theory and experiment in this episode is in direct contradiction to the results of White, Sullivan, and Barboni.[90] They examined the entire field of weak interactions during this period and concluded that theory and experiment proceeded independently. Their evidence: citations of papers, theory to theory and experiment to experiment, were higher than would be predicted from a random-citation model. It is more plausible to assume that the random model is wrong. Experimental papers might very well, and perfectly naturally, refer to all, or most, of the previous experimental work on the same subject and mention only a single theoretical paper, or perhaps a review paper, relevant to the experiment. Similarly, theoretical papers will refer extensively to other theoretical work but will cite only the most recent or best experimental paper. They might rather cite compilations of experimental results, such as the Berkeley tables, published at regular intervals. Indirect evidence supporting this view is given by White, Sullivan, and Barboni themselves. In discussing the $V-A$ theory of weak interactions, proposed independently by Feynman and Gell-Mann and by Marshak and Sudarshan to explain parity violation, they remark: "But for reasons that are not totally clear to us, Feynman and Gell-Mann seem to be given priority by physicists in the field." It is reasonable to assume that this priority reflects itself in the citations.

We see also in this episode the important role that ad hoc hypotheses play in physics. By "ad hoc," I mean a hypothesis invented specifically for the purpose of explaining a particular experimental result. All the alternative explanations offered were ad hoc in this sense, a point explicitly made by Laurent and Roos: "We want to stress the *ad hoc* character of the non-linear model for K decay that we are going to use. This model has as its only purpose to show that a CP invariant but non-linear theory can give results in agreement with the experiments."[91] Each alternative was not only experimentally testable but also tested.[92]

This seems to be a pragmatic example of the Duhem–Quine problem.[93] Duhem and Quine pointed out that the disconfirmation of an experiment E deduced from a hypothesis H does not falsify H, but rather $H \wedge B$, where B represents background knowledge and auxiliary hypotheses. One does not know where to place the blame for the failure to observe E, and thus one cannot refute H. Let H be CP conservation and E be the absence of $K_2^0 \rightarrow 2\pi$. Observation of $K_2^0 \rightarrow 2\pi$ invalidates either CP or our background knowledge. The alternative explanations were precisely attempts to alter parts of the background. Each was tested and failed, leaving H, or CP, unprotected and violated. In a practical sense, the physics community quickly solved the Duhem–Quine problem. This does not, however, answer the logical problem posed by Duhem and Quine. There certainly exist other possible alternative explanations, but they were not regarded as physically interesting.

Swetman described this episode as a Kuhnian crisis: "The reactions to the paper of Christenson et al. encompassed an enormous range of arguments and called into question many of the underlying bases of physical theory. Fundamental results of field theory, quantum mechanics, and relativity theory were critically reexamined and challenged; new particles and fields were proposed, and cosmological influences invoked."[94] This passage does not catch the spirit of the alternatives offered: their status as interesting speculations, not expressions of serious commitment. Only three of the alternatives cast doubt on fundamental results: the exponential-decay law, pions as bosons (not mentioned by Swetman), and validity of the superposition principle. The first challenged result was indirectly tested by the end of 1965. The very fact that the physics community, both specialists in weak interactions and nonspecialists, accepted CP violation even though the second and third challenged results had not been tested, argues against the existence of any crisis, Kuhnian or otherwise. Swetman also failed to note that questions concerning these discrete symmetries had been raised much earlier, both before the discovery of parity nonconservation and after.[95]

Physicists are able to resolve questions without worrying explicitly about the problems raised by philosophers of science. Furthermore, it is not clear that Kuhn's model of scientific change, which seems to have been modeled on the Copernican revolution, should be taken seriously as a description of twentieth-century high-energy physics.[96] Evidence has already been presented in Chapter 1 that in the case of parity violation, Kuhn's model of scientific revolutions, which denies that experiment can convince the entire scientific community, is wrong. The historical record clearly shows that the experiments of Wu, Lederman, and Telegdi and their collaborators provided evidence for a major, if not revolutionary, change in the theory of weak interactions. Kuhn's recent history of the origins of quantum theory,[97] certainly a revolutionary change in physics, does not draw explicitly on his theory of scientific revolutions, perhaps indicating by omission that he does not think it applies.[98] It appears that experimental evidence has become considerably more important in deciding between competing theories that it was in the sixteenth century. One of the important aspects of the scientific revolution was, after all, the increasing role of experimental evidence in choice and development of theory.

CP violation was directly established in 1967, without any possibility of plausible alternative explanation, by the observation of charge asymmetries in the decays $K_2^0 \to \pi^{\mp} \ell^{\pm} \nu$, where ℓ is either an electron or a μ meson. Letting $\delta_\ell = [R(K_2^0 \to \pi^- \ell^+ \nu) - R(K_2^0 \to \pi^+ \ell^- \nu)]/[R(K_2^0 \to \pi^- \ell^+ \nu) + R(K_2^0 \to \pi^+ \ell^- \nu)]$, Dorfan[99] found $\delta_\mu = (4.03 \pm 1.34) \times 10^{-3}$, and Bennett[100] obtained $\delta_e = (2.23 \pm 0.36) \times 10^{-3}$. The latest average gives $\delta_\mu = (3.19 \pm 0.38) \times 10^{-3}$ and $\delta_e = (3.33 \pm 0.14) \times 10^{-3}$. CP conservation requires that $\delta_\ell = 0$, because the CP operation changes $\pi^+ \ell^- \nu$ into $\pi^- \ell^+ \nu$, and their decay rates must then be equal.

There is still good experimental support for the CPT theorem in the equality of both mass and lifetime measurements of particle–antiparticle pairs. There has been, however, no direct observation of violation of time-reversal invariance, despite many attempts. Casella[101] has argued that considering the phenomenological parameters of $K_2^0 \to 2\pi$ decay and the requirements that T invariance sets on the values of these parameters, experimental results can in fact test T violation in $K_2^0 \to 2\pi$ decays, independent of the validity of the CPT theorem: " . . . *we conclude that present data suffice to establish that time-reversal symmetry is broken in the decays of the K^0 meson into two pions* [italics in original]. Our result is consistent with but otherwise independent of the validation of the CPT theorem."[102] This view has not been generally accepted by the physics community, who seem to prefer direct experimental evidence. The situation is much the same as it was in 1965. The CPT theorem has strong experimental and theoretical support; CP is violated; T therefore does not hold. But there is no direct experimental confirmation of its failure.

Professors Cronin and Fitch were awarded the Nobel Prize in physics in 1980 for their pioneering work.

4

The role of experiment

In the first three chapters we saw some of the varied roles that experiment can play in the choice between competing theories or hypotheses. In the case of parity nonconservation, the experiments of Wu, Garwin, and Telegdi and their collaborators were "crucial" experiments that decided the issue between two classes of theories: those that conserved parity and those that did not. The Princeton experiment was an example of a "convincing" experiment. It provided the evidence that persuaded the majority, though not all, of the physics community that CP invariance was violated. That episode also produced experiments refuting or eliminating most of the alternative explanations to CP violation. This was an example of a pragmatic solution to the Duhem–Quine problem.

The episode of the nondiscovery of parity nonconservation illustrated two other roles for experiment. The experiments on Mott double scattering at 0° and 180° did not lead to refutation or rejection of Dirac's theory.[1] They did lead to various unsuccessful ad hoc modifications of the theory and to repetitions of the experiment until the discrepancy was resolved. The double-scattering experiments at 90° and 270°, which, in retrospect, should have been interpreted as showing parity nonconservation, did not seem to have much effect at all. They did not lead to any new theoretical explanations, even though it was recognized at the time that existing theory provided no explanation for their results, nor did they lead to much further experimentation. Their significance was not realized at the time; they were first neglected, and then forgotten, until 1957.

These examples do not exhaust the list of roles that experiment plays in physics. Ian Hacking has discussed these and argued persuasively that experiment often has a life of its own, independent of theory.[2] Among these other roles is that of confirming or corroborating existing theory. As will be discussed later, experiment can provide us with reasons for believing a theory. In a later chapter we shall see examples of a complex feedback relationship between theoretical explanation and experimental observation. The observations help one to decide between competing theories or help to confirm a theory. The accepted theoretical explanation then provides some support for the experimental results. These will be illustrated in discussions of the discovery of synchrotron radiation from Jupiter and the discovery of the $W^{\pm}$ bosons. Experiment can also indicate the need for a new theory. The experiments on parity violation clearly showed the need for a new theory (all existing theories at that time conserved parity) and led to the $V-A$ theory of weak interactions.

Most experiments fall into the class of experiments that measure quantities.[3] These are not independent of theory, because theory often determines what quantities are of interest. They are not, however, regarded as means of testing a theory. Thus, we find considerable efforts devoted to spectroscopy both in the nineteenth century, before any theory of spectra existed, and in the twentieth century, when first Bohr theory and then later quantum mechanics explained the findings and increased their importance. Andy Pickering has remarked on this in his discussion of contemporary high-energy physics experiments.[4] He remarks that under the guidance of currently accepted theory, high-energy physics experiments look at rather rare phenomena and analyze them in ways that are not obvious, in contrast to the 1960s, when scattering and production experiments, which were easy to observe, dominated experiment. Pickering regards this effect of theory on experiment as harmful. Although it does seem to have an unfortunate aspect, namely, the fact that exploratory experiments, in which physicists simply want to look at some aspect of nature to see what is going on, are becoming very rare in high-energy physics, this is not the case in all fields. Such experiments are often carried out in the fields of condensed matter or cosmic-ray physics. In addition, experiments in high-energy physics require so much space and are so expensive, and time at the large accelerators so limited, that without a strong theoretical justification such experiments are unlikely to be funded or scheduled. What Pickering overlooks is that theory has always provided guidance for experiment. We do not examine all of nature, but rather devote our attention to what seem to be the most important aspects of it. At least one guide to what is important is theory. It is only in retrospect that the scattering experiments and those that looked for peaks in the invariant-mass plots appear intuitive and obvious. What does seem to be a legitimate worry is that somehow this guidance will become dominance: Effects that are not predicted by theory will not be looked for or perhaps not be perceived even when the data are there. An examination of only the episodes presented here will serve to reduce this worry. After all, both parity and CP violation, which not only were not predicted by existing theory but also were in conflict with the strongly held beliefs of the physics community, were observed, as were the anomalies attending the well-corroborated Dirac theory. These issues will be discussed in detail in Chapter 5, where the question whether or not the preconceptions of the experimenter influence the experimental results is discussed, and in Chapter 8, where the question of theoretical or experimental bandwagons, in which previous results or existing theory determine observations, will be considered.

In this chapter I wish to comment on the role that experiment plays in theory choice and in the confirmation of theory. The discussion in the previous chapters has been primarily descriptive. I now wish to discuss some of the questions that philosophers of science have raised concerning this role. These will include the Duhem–Quine problem, mentioned earlier, the issue of the theory-ladenness of observation and the related problem of incommensurability, and the Bayesian approach to confirmation of theories. This will certainly not be a complete or exhaustive treatment of these questions. I do hope to show, however, as has been indicated earlier, that the actual practice of physics is relevant to discussions of these issues. The historical episodes also show that scientists do seem to attempt pragmatic solutions to these problems, even when they may not be aware of doing so. I suggest that it might be valuable for appraisals of scientific work if these discussions were more explicit. Thus, the philosophy of science may be of some relevance to the practice of science.

4.1 THE DUHEM–QUINE PROBLEM[5]

The Duhem–Quine problem can be summarized briefly as follows: In the usual *modus tollens*, if a hypothesis H implies an experimental result E, then failure to observe E, or $\neg E$, implies $\neg H$. Duhem and Quine pointed out that it is never simply H alone that implies E, but rather $H \wedge B$, where B includes background knowledge and auxiliary assumptions. Thus, $\neg E$ implies $\neg H$ or $\neg B$, and we are never quite sure which part of our conjunction has been refuted. We may then be able to save H from refutation provided we are willing to make changes in the background knowledge or auxiliary assumptions. As Quine put it, "Any statement can be held true come what may, if we make drastic enough adjustments elsewhere in the system."[6] In a sense, it is only science as a whole that faces experiment, not particular theories or hypotheses.

I should emphasize here that I am dealing with a weak form of the Duhem–Quine problem. I am assuming that the experimental result $\neg E$ is correct. The strong form, in my view, allows the experimental result itself to be questioned.[7] I recognize, of course, that experimental results are fallible and that we might have to change our judgment later. I shall, however, argue in a later chapter that we are always given good reasons for belief in a particular observation. A devout skeptic may nevertheless question the result. During a discussion of the parity nonconservation episode I was asked how I knew that something somewhere else in the universe did not invert just as the experiment of Wu was being done. The answer, of course, is that I cannot know this, although in this case the nonconservation of parity locally would be of considerable interest. There is, I believe, no answer to this type of skepticism. It just seems irrelevant to science.

Our previous discussion indicates that for the physics community, this weaker form of the problem can be solved. In discussing the episode of parity nonconservation I have argued that the decision is between classes of theories, those that do and those that do not conserve parity, and that these classes are both exclusive and exhaustive. Thus, the hypothesis "parity is conserved in the weak interactions" was both tested and refuted. In the case of CP violation, we saw a pragmatic solution of this problem in which several alternative explanations to CP violation were offered, most of which were eliminated by experiment. Arguments were presented to suggest why these alternative explanations were offered. These included the complexity of the K^0 system and the lack of complete understanding of the system, the smallness of the observed effect (in absolute, although not statistical, terms), the fact that this effect involved a new and very small energy region that previously had been unexplored, and the unavailability of the Landau solution, which tied parity violation to the existence of a real structure in the world (particles and anti-particles).

A worrisome question still remains, however. The argument in the case of parity violation, in which the two classes of theories are both exclusive and exhaustive, would seem to apply just as well to the episode of CP violation. After all, there are only two classes of theories: those that conserve CP and those that do not. What is the difference between the two episodes? I believe the difference lies in the length and complexity of the derivation linking the hypothesis to the experimental result. In the case of parity, the experiments could be seen by inspection to violate mirror symmetry (see Figure 1.2). In the CP episode, what was observed was $K_2^0 \to \pi^+ \pi^-$. In order to relate this observation to the question of CP violation, one had to assume (1) the principle of superposition, so that the K_1^0, K_2^0 description could be used, (2) that the exponential-decay law held for K mesons out to 300 lifetimes, so that one could conclude that it was K_2^0 decay being observed, (3) that the decay particles were both "real" pions and that pions were bosons, (4) that no other similar particle was produced, and (5) that there were no external conditions present that might regenerate K_1^0 mesons. This is a far cry from the rather simple derivation in the parity experiments, and these assumptions were tested and eliminated by subsequent experiments (see discussion in Chapter 3).

Another interesting question suggested by these episodes is why the oft-repeated and well-documented discrepancies between Mott's theoretical calculation and the experimental results on double scattering of electrons did not result in refutation of Dirac's electron theory. Here, too, the derivation of the effect involved several assumptions, including single scattering, at high velocity, off high-Z nuclei at large angles and the assumption that other effects such as screening of the nucleus by atomic electrons, multiple scattering, and modifications of Coulomb's Law were negligible. As we have seen, the discrepancy was worrisome, and questions were raised whether or not the experimental apparatus satisfied the conditions for Mott scattering and whether or not these other effects might be present. I would also suggest that not all experiments have equal weight in testing a theory. Dirac's theory had had important successes in predicting the existence of the positron and in atomic spectroscopy; the discrepancies in electron scattering and in the spectrum of hydrogen[8] simply did not carry much weight in comparison with these successes. It appears that we are dealing with something like a preponderance of evidence.[9] I suggest that the fact that the positron was not predicted by any other existing theory and was quite unexpected on the basis of background knowledge added to its importance. Perhaps if there had been a competing theory that had successfully explained the discrepancies, the situation might have been different. The Bayesian approach to confirmation theory, to be discussed later, does provide a natural explanation of this episode and a solution to this particular case of the Duhem–Quine problem.

The discussion of Duhem and Quine emphasizes a particular problem in the role of experiment in testing theories. As will be discussed later, physicists tend to test locally those parts of a theory that are deemed crucial for derivation of the effect. They do not seem to test general theories very often. Clark Glymour has offered an interesting solution to this problem in his "bootstrap" method, in which the particular statements needed for the derivation are isolated.[10] This does not seem to fit many of the cases of localized testing. Thus, in the early days of the V–A theory of weak interactions, when the predicted ratio of $\pi^+ \rightarrow e^+ \nu / \pi^+ \rightarrow \mu^+ \nu = 10^{-4}$ was not observed,[11] it was the use of the vector and axial-vector interactions that was questioned, not quantum-field theory, which was also needed for calculation of the result. A similar example concerns light scattering from CdS crystals. In that case, two different calculations, one that included excitons and one that did not, were compared. The authors concluded that excitons were important: "This demonstration of exciton resonance further supports the theory of Ganguly and Birman."[12] An alternative view, that of modifying quantum mechanics, was nowhere mentioned. As the principal investigator, James Scott, remarked, "Such a suggestion would have been laughed at."[13]

I know of no general solution to this problem. Glymour's approach seems to be able to localize the problem to those statements necessary for derivation of the experimental result, but goes no further. In the cases mentioned earlier, both the general theory and the specific auxiliary hypothesis were needed. Jon Dorling has offered a Bayesian solution[14] to the Duhem–Quine problem. He has shown that under a wide variety of conditions, the decision to reject the auxiliary hypothesis rather than the general theory can be justified, even when the auxiliary hypothesis is more strongly believed initially. This is certainly a major step toward a solution, but questions remain concerning its general applicability and whether or not the conclusions depend on the initial degrees of belief.[15]

What can we conclude from this discussion? It is, of course, dangerous to generalize from only a few cases, no matter how clear or exemplary. Nevertheless, the fact that there exists at least one "crucial" experiment, parity violation, that can be philosophically justified, and the "convincing" CP-violation experiment argue against those who would deny or minimize the role of experiment in theory choice. Parity violation may show that the Duhem–Quine statement that no single hypothesis can ever be tested alone is false in at least the weak sense we have discussed, but that is not the point. These episodes show that the physics community, at least implicitly, has tried to solve this problem, with varying success. What we need are more detailed studies to help clarify the issue. I doubt that there is a general answer to the question what role experiment does and should play in theory choice, but the discussion is surely valuable.

4.2   THE THEORY-LADENNESS OF OBSERVATION AND INCOMMENSURABILITY

Many writers[16] have remarked on the theory-ladenness of observation, the fact that terms that are used in observation statements or measurement reports are laden with a particular theory. For example, although terms like "force" and "mass" appear in observation statements, they take their meanings only in the context of Newtonian dynamics. The question then arises whether or not such theory-laden terms can be used in experiments that test theories. One should be careful here to distinguish, if possible, between the theory of the experimental apparatus and the theory under test. If they are distinct, no major problems arise. We noted earlier that at least part of the argument concerning the experiments on parity violation was that the theory of the apparatus was quite different from the theory of the weak interactions, which was being tested. That clearly would not be the case if one used a mercury thermometer to investigate whether or not objects expanded as their temperatures rose. The operation of the apparatus would depend on the hypothesis being tested.

What happens when the theory of the apparatus and the theory of the phenomena are not distinct? Thomas Kuhn and others, including Paul Feyerabend and Barry Barnes, have argued that there can be no comparison of competing paradigms based solely on experimental evidence: "There is no appropriate scale available with which to weigh the merits of alternative paradigms: they are incommensurable."[17] Briefly stated, the argument is as follows: There can be no neutral observation language. All observation terms are theory-laden, and thus we cannot compare experimental results, because in different paradigms, terms describing experimental results have different meanings, even when the words used are the same.[18] An example would be the term "mass," which in Newtonian mechanics is a constant, whereas in Einstein's special theory of relativity it depends on velocity. I believe this view is incorrect. We discussed earlier the experiments on double scattering of electrons that tested and disagreed with Dirac's electron theory, even though the meaning of "electron" was, at least in part, determined by that theory. We shall also see later that the theory-ladenness of experimental apparatus can be of assistance in discussing confirmation of theories or hypotheses.

In this discussion I wish to consider a more abstract, "invented" experiment that provides a clear discussion of the issue. I shall demonstrate that an experiment, described in procedural, theory-neutral (between the two competing paradigms or theories) terms, gives different theory-neutral results when interpreted in the two alternative paradigms. Thus, a measurement of the quantity derived will unambiguously distinguish between the two. I shall take as an example one of Kuhn's own exemplars: the difference between Newtonian and Einsteinian mechanics. The case in point can be loosely described as the scattering of equal-"mass" objects.

The experimental procedure is as follows: Consider a class of objects, let us say billiard balls.[19] The objects are examined pairwise by placing a compressed spring between them. The spring is allowed to expand freely, and the velocities of the two objects are measured. Because we restrict ourselves to a single frame of reference in the laboratory, the measurement is theory-neutral. We then select two balls whose velocities are equal. Of course, a Newtonian would interpret this as giving two objects for which the Newtonian masses M_N are both equal and constant, whereas an Einsteinian would say that the relativistic masses $M_R = M_{0R}/\sqrt{1-v^2/c^2}$ are equal and thus that the rest masses M_{0R} are equal. This is agreed, but the point is that the procedure itself is theory-neutral. One of the objects is placed at rest in the laboratory, and the other is given a velocity V_1 (again, theory-neutral), and the objects are allowed to scatter off each other. Care is taken to make the collision elastic (i.e., no energy of any sort is emitted). The final velocities of the objects, V_2 and V_3, are measured, as is the angle between these velocities. It is again true that the assignments of momenta and energies to the particles will be different, but measurement of the angle does not depend on these assignments. As is well known (and as will be demonstrated later), the predicted value of this angle differs in Newtonian and relativistic mechanics. For a Newtonian, $\theta_N = 90°$, whereas for an Einsteinian, $\theta_R < 90°$. In the following calculation we shall use units such that $c = 1$ (the velocity of light). We shall also assume for calculational convenience that the two outgoing velocities are equal. This apparent loss of generality is unimportant, although the proof can be generalized.[20] The fact that any single experiment can distinguish between the two paradigms is enough to show that they are commensurable.

Let us first consider the experiment from a Newtonian point of view. Let $\mathbf{P}_{1N}$, $\mathbf{P}_{2N}$, $\mathbf{P}_{3N}$ be the initial and final momenta and E_{1N}, E_{2N}, E_{3N} be the initial and final energies. We further note that in this case $E_{1N} = P_{1N}^2/2M_N$. We now apply the Newtonian law of conservation of momentum:

$$\mathbf{P}_{1N} = \mathbf{P}_{2N} + \mathbf{P}_{3N}$$

$$P_{1N}^2 = P_{2N}^2 + P_{3N}^2 + 2P_{2N}P_{3N}\cos\theta_N$$

That gives

$$\cos\theta_N = (P_{1N}^2 - P_{2N}^2 - P_{3N}^2)/2P_{2N}P_{3N} \tag{4.1}$$

Applying Newtonian conservation of energy, we have $E_{1N} = E_{2N} + E_{3N}$, or

$$P_{1N}^2/2M_N = P_{2N}^2/2M_N + P_{3N}^2/2M_N$$

That gives

$$P_{1N}^2 - P_{2N}^2 - P_{3N}^2 = 0 \tag{4.2}$$

Combining equations (4.1) and (4.2), we get the result

$$\cos \theta_N = 0 \quad \text{or} \quad \theta_N = 90°, \quad \text{because } P_{2N} \text{ and } P_{3N} \neq 0$$

Let us now consider the experiment interpreted in relativistic terms. Using relativistic conservation of momentum, we have $\mathbf{P}_{1R} = \mathbf{P}_{2R} + \mathbf{P}_{3R}$, and we obtain, by a calculation identical with that earlier,

$$\cos \theta_R = (P_{1R}^2 - P_{2R}^2 - P_{3R}^2)/2P_{2R}P_{3R}$$

We can rewrite this, using relativistic energy and momentum, as

$$\cos \theta_R = (E_{1R}^2 - M_{0R}^2 - E_{2R}^2 + M_{0R}^2 - E_{3R}^2 + M_{0R}^2)/2P_{2R}P_{3R}$$

$$\cos \theta_R = (E_{1R}^2 - E_{2R}^2 - E_{3R}^2 + M_{0R}^2)/2P_{2R}P_{3R} \tag{4.3}$$

Using relativistic conservation of energy, we have

$$E_{1R} + M_{0R} = E_{2R} + E_{3R} \tag{4.4}$$

Squaring both sides gives

$$E_{1R}^2 + 2E_{1R}M_{0R} + M_{0R}^2 = E_{2R}^2 + 2E_{2R}E_{3R} + E_{3R}^2 \tag{4.5}$$

Rewriting, we have

$$E_{1R}^2 - E_{2R}^2 - E_{3R}^2 + M_{0R}^2 = 2E_{2R}E_{3R} - 2E_{1R}M_{0R} \tag{4.6}$$

We can now show that the expressions in equation (4.6), which compose the numerator in the expression for $\cos \theta_R$ [equation (4.3)], are greater than zero and that $\cos \theta_R > 0$, or $\theta_R < 90°$. Consider $2E_{2R}E_{3R} - 2E_{1R} M_{0R}$. We now consider the case in which $E_{2R} = E_{3R} = (E_{1R} + M_{0R})/2$, from equation (4.4):

$$2[(E_{1R} + M_{0R})/2]^2 - 2E_{1R}M_{0R}$$
$$= 2(E_{1R}^2 + 2E_{1R}M_{0R} + M_{0R}^2)/4 - 2E_{1R}M_{0R}$$
$$= E_{1R}^2/2 - E_{1R}M_{0R} + M_{0R}^2/2 = [(E_{1R} - M_{0R})/\sqrt{2}]^2$$

By the condition of our experiment, $E_{1R} > M$, and thus this expression is greater than zero.

Thus, $2E_{2R}E_{3R} - 2E_{1R}M_{0R} = [(E_{1R} - M_{0R})/\sqrt{2}]^2$ in this case, but $2E_{2R}E_{3R} - 2E_{1R}M_{0R} = E_{1R}^2 - E_{2R}^2 - E_{3R}^2 + M_{0R}^2$ and must be greater than zero, and hence $\cos \theta_R > 0$ and $\theta_R < 90°$, because $P_{2N} = P_{3R} \neq 0$.

We have shown that in this procedurally defined, theory-neutral experiment, the predicted results, when interpreted within the two paradigms, are unambiguous and different; that is, $\theta_N = 90°$, $\theta_R < 90°$. A measurement of the angle between the velocities of the two outgoing objects will clearly distinguish between the two paradigms, and the paradigms are commensurable. The fact that two paradigms are commensurable does not require that a scientist accept the one consistent with the experimental results, namely, relativity. A clever classical physicist might very well accommodate the results by postulating new mechanical effects depending on motion through the ether. These two issues of commensurability and theory choice are often conflated, as exemplified in the statement by Barnes cited earlier, but it is useful to keep them separate and distinct. The point here is that, in this case, even though the experimental terms are theory-laden, one can describe an experiment in relatively theory-neutral terms, so that the theories can be both distinguished and tested. This is not, of course, a general proof that such experiments can always be constructed. It should be pointed out, too, that the experiments of Lummer and Pringsheim and of Rubens and Kurlbaum on the spectrum of black-body radiation, which led to Planck's suggestion of quantization, also seem to be neutral with respect to classical and quantum physics. If this type of experiment can be found for what are usually regarded as two of the most revolutionary changes in physics, then I suspect they can be found for almost all cases. However plausible the idea of the theory-ladenness of observation is, it has not been demonstrated that it prohibits theory testing. This discussion has shown that one of Kuhn's favorite and most important examples is incorrect, and I know of no better illustration of incommensurability.

4.3   BAYESIAN CONFIRMATION THEORY

In Section 4.1 I suggested that the Bayesian approach to confirmation theory could explain, in a rather natural way, the episode in which failure to observe the predicted asymmetry in double scattering of electrons did not lead to refutation of Dirac's theory, but rather to attempts to modify the auxiliary hypotheses, a pragmatic solution of the Duhem–Quine problem. In this section I shall comment on this Bayesian approach[21] and its success in explicating both scientific method[22] and the role that experiment plays in confirmation of theories or hypotheses. I shall also demonstrate later how this method can explain the Dirac episode. This is not to say that scientists are conscious Bayesians, but rather that the Bayesian approach provides a description of what happens in various scientific episodes. This approach certainly has its critics, and we shall discuss some of the criticisms and objections later.

The basis of Bayesianism, as it is called, is that we have degrees of belief in statements or hypotheses and that these degrees of belief obey the probability calculus. There has been much discussion as to what kind of probabilities these are.[23] At the moment there seems to be no good way to make these probabilities logical or objective,[24] and I believe that the best we can do is to settle for subjective probabilities, or betting behavior. The underlying mathematical statement of this approach is known as Bayes's Theorem. It states that

$$P(h \wedge e) = P(h \mid e)P(e) = P(e \mid h)P(h) \tag{4.7}$$

where $P(h \mid e)$ is the conditional probability of h given e, and $P(e)$ and $P(h)$ are the prior probabilities of e and h, respectively. These probabilities are all assumed to be relativized to some background knowledge.

One immediate objection to the Bayesian approach is that if the prior probabilities, representing our degrees of belief, are subjective, then it is possible for different scientists to have very different prior probability assignments and thus to disagree about $P(h \mid e)$. Savage has shown,[25] however, that provided the prior probability assignments for a particular hypothesis are neither 0 nor 1, then Bayesian conditionalization will eventually bring two observers into close agreement about the significance of the available evidence, no matter how different their initial probability assignments were. The formal result does involve an indefinite amount of available evidence to induce this convergence, but in actual science this does not seem to be a significant problem. Scientists do seem to converge to a single view on the basis of quite finite evidence. A related criticism is that the Bayesian approach offers no method for assigning prior probabilities to a set of competing hypotheses. This is similar to criticizing deductive logic because it has no method of assigning truth values to its premises.[26] Bayesianism provides a method for changing our beliefs in the light of evidence. It need not say what those beliefs should be. In light of Savage's result, discussed earlier, all that is needed is some reasonable way of assigning finite probabilities to the hypotheses, as, for example, giving each of n competitors a prior probability of $1/n$. Other schemes are, of course, possible.[27]

One reason for adopting the Bayesian approach is that it gives us the intuitively appealing result that observation of the evidence entailed by a hypothesis h increases our degree of belief in h. Rewriting equation (4.7), we obtain

$$P(h \mid e) = \frac{P(e \mid h)P(h)}{P(e)} \tag{4.8}$$

When h entails e, $P(e \mid h) = 1$. Thus, $P(h \mid e) = P(h)/P(e)$, so that for $P(e) < 1$, which is surely the case almost all of the time, $P(h \mid e) > P(h)$.

This would seem to be a requirement for rational behavior. If observation of evidence entailed by a hypothesis does not increase our degree of belief in that hypothesis, what should?

We are now in a position to illustrate the value and fruitfulness of the Bayesian approach by demonstrating that it can explicate, and quite naturally, the episode concerning Dirac theory and the results on double scattering of electrons mentioned earlier and discussed in detail in Chapter 2.

Let D be the Dirac Theory, and let e_1 be the existence of the positron. Applying Bayes's Theorem, we get

$$P(D \mid e_1) = \frac{P(e_1 \mid D)P(D)}{P(e)}$$

where $P(e_1 \mid D) = 1$, because $D \vdash e_1$. We can also write $P(e_1) = P(e_1 \mid D)P(D) + P(e_1 \mid \neg D)P(\neg D)$, where $\neg D$ indicates negation of D. Because no other theory, at the time, predicted the existence of the positron, we can set $P(e_1 \mid \neg D)P(\neg D) = \epsilon$, where ϵ is very small. Thus,

$$P(D \mid e_1) = \frac{P(D)}{P(D) + \epsilon} \approx 1 - \epsilon \approx 1 \tag{4.9}$$

This confirms our intuition that observation of a result predicted by only one theory gives us very good reason to believe that theory.

We now set $P(D \mid e_1) = P_1(D)$, the new prior probability of D. Let us now consider what happens to our belief in D when e_2, symmetric double scattering of electrons, is observed:

$$P(D \mid e_2) = \frac{P(e_2 \mid D)P_1(D)}{P(e_2)} \tag{4.10}$$

$$P(e_2) = P(e_2 \mid D)P_1(D) + P(e_2 \mid \neg D)P_1(\neg D) \tag{4.11}$$

Let $K = P(e_2 \mid \neg D)/P(e_2 \mid D)$; then, from equation (4.11),

$$P(e_2) = P(e_2 \mid D)P_1(D) + KP(e_2 \mid D)[1 - P_1(D)] \tag{4.12}$$

but $P_1(D) = 1 - \epsilon$,

$$P(e_2) = P(e_2) \mid D)[1 - \epsilon] + KP(e_2 \mid D)[1 - 1 + \epsilon] \tag{4.13}$$

$$P(e_2) = P(e_2 \mid D)[1 - \epsilon + K\epsilon]$$

We can neglect ϵ, which is very small; so, from equation (4.10),

$$P(D \mid e_2) = \frac{P(e_2 \mid D)P_1(D)}{P(e_2 \mid D)[1 + K\epsilon]} = \frac{P_1(D)}{1 + K\epsilon} \tag{4.14}$$

We expect $K \leq 1$, because $\neg D$ is a weak statement compared with D, and in this case we see $P(D \mid e_2) \approx P_1(D)$, or our confidence in D remains virtually unchanged. The result is quite robust for differing values of K, depending on ϵ. Thus, if $\epsilon = 0.01$, K can be as large as ten without changing our belief in D by more than 10 percent.

Let us now see what we can say about $P(H \mid e_2)$, where H represents the auxiliary hypotheses used by Mott to derive $\neg e_2$, the asymmetric scattering: single, large-angle scattering of high-velocity electrons from a heavy nucleus. By a theorem of the probability calculus,

$$P(H \mid e_2) + P(D \mid e_2) - P(H \wedge D \mid e_2) = P(H \vee D \mid e_2) \tag{4.15}$$

but $P(H \wedge D \mid e_2) = 0$, because $H \wedge D \vdash \neg e_2$; so

$$P(H \mid e_2) = P(H \vee D \mid e_2) - P(D \mid e_2) \tag{4.16}$$

But $P(H \vee D \mid e_2) \geq P(D \mid e_2)$, and so we can write $P(H \vee D \mid e_2)$ as $1 - \delta$, where $\delta < \epsilon$. $P(H \mid e_2) = 1 - \delta - (1 - K\epsilon) = K\epsilon - \delta$, which is very small.

Thus, as both our simple calculation and the history show, our belief in H becomes very small, while our belief in D remains virtually the same. As we have seen, scientists did offer alternatives to H, and the final explanation of the failure to see asymmetric double scattering was the larger effect of plural scattering.

This is a solution to a very particular case of the Duhem–Quine problem in which $P(D)$ is originally very large. In the general case, in which $P(D \mid e_2)$ is not very close to 1, we cannot say anything, as equation (4.16) shows. It does indicate, however, that when $P(H \mid e_2)$ is close to 1, D will receive the blame, as we expect.

This derivation also shows the value of the Bayesian approach, because it can explain an episode that otherwise is explicable only in the vaguest sense. Other examples of this value will be given later.

Two problems of Bayesianism[28]

There are two problems that have been, quite legitimately, regarded as difficulties for the Bayesian view. The first of these is the "tacking paradox" or the problem of localization of support.[29] Let us assume that a sufficient condition of support of h by e is that h entails e. Then, if h entails e, so does $h \wedge h'$, and so, according to this view, $h \wedge h'$ is also supported by e, for arbitrary h'. This seems somewhat counterintuitive. If "quantum mechanics" entails the Balmer series, then so does "quantum mechanics" with the addition of the statement "The moon is made of green cheese." The Bayesian answer to this is as follows.

Let us assume the usual Bayesian support function $S(h, e) = P(h \mid e) - P(h)$. For the case $t \vdash h \vdash e$ we get

$$P(t \mid e) = \frac{P(e \mid t)P(t)}{P(e)} \begin{cases} \text{but } P(e \mid t) = 1 \\ \text{and } P(t) = P(t \mid h)P(h) \end{cases}$$

$$= \frac{P(t \mid h)P(h)}{P(e)}$$

and

$$S(t, e) = P(t \mid e) - P(t)$$

$$= \frac{P(t \mid h)P(h)}{P(e)} - P(t \mid h)P(h)$$

$$= P(t \mid h)\left(\frac{P(h)}{P(e)} - P(h)\right)$$

$$= P(t \mid h)[P(h \mid e) - P(h)]$$

$$= P(t \mid h)S(h, e)$$

One can further argue that the excess content of t over h is, in fact, given by $-\ln P(t \mid h)$. Thus, the support of t by e, as compared with the support of h by e, is diluted by an increasing function of the excess content of t over h, which is intuitively appealing. The more the content of a general theory t goes beyond that of h, the specific hypothesis that is needed to derive e, the less it is supported by e.

In the case mentioned earlier, let h = "quantum mechanics," let h' = "The moon is made of green cheese," and let $t = h \wedge h'$. Then $P(t \mid h) = P(h' \mid h) = P(h')$, because h' is independent of h. Thus, $S(t, e) = P(h') \, S(h, e)$, which is considerably reduced by the probability of belief in the statement concerning the moon's composition. Because the moon is known not to be made of green cheese, $P(h') = 0$, and thus the conjunction, in fact, receives no support.

The second problem is that of known, or old, evidence. If evidence e is known, then $P(e) = 1$, and from the foregoing, $P(h \mid e) = P(h)$, and h receives no support from e. This seems to be at variance with the history of science, for there have been numerous cases in which a theory or hypothesis has been invented for the purpose of solving a known problem or explaining known data, and the solution or explanation has counted as support for the theory. A particularly telling example of this will be discussed later. I believe that adequate Bayesian solutions have been provided by Howson,[30] Garber,[31] and Niiniluoto.[32] Howson and Garber point out that known facts can support a hypothesis because the correct background knowledge to be used should exclude those facts. Niiniluoto suggests that facts known and used in the construction of a hypothesis increase the prior probability of the hypothesis. Known facts that are not initially believed to be relevant to the hypothesis do, in fact, support it.

This view is at variance with a widely held non-Bayesian position, due primarily to Zahar[33] and Worrall,[34] in which facts both known and used in the construction of a hypothesis cannot be used to support it. This position not only disagrees with the history of science but also gives a rather odd result in the following example. Let us consider an urn containing 100 balls, from which the balls are removed one by one and examined for color. All of the balls are found to be green. Scientist A then formulates hypothesis h, "All the balls in the urn are green." Scientist B, somewhat more impetuously, formulated hypothesis h', "All the balls in the urn are green," before the examination of the balls. Surely, despite our whimsical notation, these are the same hypothesis and must receive the same support by the evidence. In fact, the hypothesis receives maximal support by the evidence, because it is entailed by the evidence. What, then, is the difficulty with the Zahar–Worrall position? It is that they are conflating credit to the scientist with confirmation of the hypothesis. Scientist A receives no credit for h,[35] but h is confirmed. If scientist B had formulated h on the basis of some theory t, then B would receive credit, and t would receive diluted support, as shown earlier. In addition, they are trying to guard against content-reducing stratagems or exception-barring. This seems to be a problem only in the examples given by philosophers of science (i.e., "All bread, except that of a village in France, nourishes"), not in science itself (see discussion in Chapter 3). We may also ask what is necessarily wrong with a content-reducing stratagem. If we have overgeneralized, such a tactic might be the only way of correcting our mistake.

I believe the position of Zahar and Worrall to be wrong, not only because of the work and arguments cited earlier but also because it is in conflict with a well-known and extremely relevant historical example. This is the case involving Kepler's Laws and Newton's Inverse-Square Law of Universal Gravitation. It seems quite clear from both historical and modern evidence[36] that Kepler's Laws, which were certainly known to Newton, are regarded as confirming at least the inverse-square segment of the gravitational-force law.[37] This episode differs from the cases already considered by Howson and Niiniluoto, because Kepler's Laws were used to derive, in at least an idealized or approximate way, Newton's Inverse-Square Law of Gravitation. It can be easily accommodated within a Bayesian account of confirmation. Later I shall consider the philosophical objections, raised by Popper, Duhem, and Zahar, that Kepler's Laws cannot confirm Newton's Law because they are contradictory; that is, universal gravitation implies that planetary orbits are not ellipses because of the perturbations due to other planets.

Newton and Kepler

Let us examine Newton's argument for his inverse-square law of gravitational force in the *Principia*.[38] He begins by describing the background knowledge needed: "It is enough if one carefully reads the Definitions, the Laws of Motion, and the first three sections of the first Book. He may then pass on to this Book, and consult such of the remaining Propositions of the first two Books, as the references in this, and his occasions, shall require" (p. 397).[39] He then describes various phenomena, empirically known, that are to be used in his derivation. He includes

That the fixed stars being at rest, the periodic times of the five primary planets (Mercury, Venus, Mars, Jupiter, and Saturn), and (whether of the sun about the earth, or) of the earth about the sun, are as the 3/2th power of their mean distances from the sun. [Phenomenon IV, Book III, p. 404]

Similar statements are made about the moons of Jupiter and Saturn. Newton regards these as established by observation: "This proportion, first observed by Kepler, is now received by all astronomers" (p. 404). He then proceeds to derive the inverse-square law for planets:

That the forces by which the primary planets are continually drawn off from rectilinear motions, and retained in their proper orbits, tend to the sun; and are inversely as the squares of the distances of the places of the planets from the sun's centre. [Proposition II, Theorem II, Book III]

The former part of the Proposition is manifest from Phen. V [Equal Area Law] and Prop. II, Book I [which proves the Equal Area Law holds for a central force]; the latter from Phen. IV [cited above] and Cor. VI, Prop. IV, of the same book. [p. 406][40]

How would a Bayesian deal with this? Let us assume as background, as Newton did, the laws of motion as well as the theorems proved for various types of motion. Let us regard as evidence, e, Kepler's Laws, empirically established, and let h be the hypothesis of the inverse-square law of force between the sun and each of the planets. Then, by Bayes's Theorem, with all probabilities relativized to our background,

$$P(h \mid e) = P(e \mid h)P(h)/P(e)$$

but $P(e \mid h) = 1$ because h entails e; so

$$P(h \mid e) = P(h)/P(e)$$

But, as demonstrated by Newton, relative to our background knowledge, we can derive h from e. Thus, $P(h) = P(e)$ and we obtain $P(h \mid e) = 1$, or maximal support for h by e, a not very surprising result if it is indeed a derivation.

One may then ask, if this is the case, why Newton presented further evidence from the motion of the moon and the moons of Jupiter and Saturn. It is here that our previous caution concerning separation of the inverse-square force from its universality should be borne in mind. The additional evidence supports the hypothesis that the inverse-square force is universal.

The careful reader may think that the foregoing result is too strong. How can it be, if $P(h \mid e) = 1$, that the discovery of Neptune was, and is, regarded as a further confirmation of Newton's Law? The point is quite simple. Herschel's observations of the discrepancy in the motion of Uranus cast doubt on the validity of Kepler's Laws and also on the derivation from those laws. Thus, $P(h \mid e')$, where e' includes the discrepancy, is clearly less than 1. When the discovery of Neptune showed that the error was in assuming that all planets were known, Newton's Law could then acquire additional confirmation, as both history and intuition would suggest.

Duhem, Popper, and Zahar[41] have objected that Kepler's Laws cannot be used to derive Newton's Law of Gravitation because they are contradictory. The problem can be separated into two parts. Even if the solar system consisted of only the sun and one planet, Kepler's Laws would be false, because Newton's Third Law of Motion (action = reaction) requires that both the sun and the planet move about their common center of gravity, whereas Kepler's Laws require a fixed sun. The second point is that if gravitation is universal (i.e., the planets and the sun mutually attract one another), then the orbits of the planets are not ellipses, and Kepler's Laws do not hold exactly because of perturbations.

Let us consider how Newton dealt with this problem. As early as 1684, in Version III of *De Motu*, he remarked as follows:

For if the common center of gravity is calculated for any position of the planets it either falls in the body of the Sun or will be very close to it. By reason of this deviation of the Sun from the center of gravity the centripetal force does not always tend to that immobile center, and hence the planets move neither exactly in ellipses nor revolve twice in the same orbit. So that there are as many orbits of any one planet as it has revolutions, as in the motion of the Moon, and the orbit of any one planet depends on the combined motion of all the planets, not to mention the action of all these on each other. But to consider simultaneously all these causes of motion and to define these motions by exact laws allowing of convenient calculation exceeds, unless I am mistaken, the force of any human mind. Ignore these *minutiae* [emphasis added], and the simple orbit and the mean among all errors will be the ellipse of which I have already treated.[42]

The discussion in Book III of the *Principia* is more detailed. Newton first considered the problem of the center of gravity. In Proposition LVIII, Theorem XXI, Book I (p. 165), he wrote

If two bodies attract each other with forces of any kind, and revolve about the common centre of gravity: I say, that, by the same forces, there may be described round either body unmoved a figure similar and equal to the figures which the bodies so moving describe round each other.

Corollary II (p.166) related the inverse-square force to orbits that were conic sections about both the center of gravity and either body, and Corollary III did the same for the equal-areas law. Although Kepler regarded his laws as referring to a fixed sun, Newton demonstrated that the first two also apply to motion about the center of gravity.

Newton was also quite aware of the problem posed by perturbations caused by the mutual interaction of the planets. In December 1684,[43] he wrote to Flamsteed requesting information concerning the motion of Saturn when it is in conjunction with Jupiter. Flamsteed supplied the information and interpreted the request as implying that Newton thought the two planets might attract each other and influence each other's motion. This is confirmed by Newton's response.

In Proposition XIII, Book III (pp. 420–1), Newton discussed the numerical value of the perturbation of Saturn's motion by Jupiter and noted that "the greatest error in the mean motion scarcely exceeds two minutes yearly"(p. 421). Similar numerical estimates for perturbations in the moon's motion about the earth, due to the sun, were given in Proposition III, Book III (p. 406).

Newton had, in fact, done more. He had considered the general problem of perturbations in a group of bodies with a mutual, inverse-square interaction. He noted that the planetary orbits would be ellipses and would obey the equal-areas law if the sun were fixed and there were no interactions: "But the actions of the planets upon one another are so very small, that they may be neglected; and by Prop. LXVI, Book I, they disturb the motions of the planets about the sun in motion, less than if those motions were performed about the sun at rest" (p. 421). He had also shown that such perturbations were minimized in a system such as our solar system, where several lesser bodies revolve about a very great one (pp. 171–3, 190).

There is one further problem in Newton's derivation. He proved the inverse-square law using Corollary VI, Proposition IV, Book I, which applies only to circular motion. Newton knew that the planetary orbits were ellipses, but regarded the deviations as small: "... in laying down these Phenomena, I neglect these small and inconsiderable errors" (p. 405).

It seems clear that Newton's derivation of the inverse-square law from Kepler's Laws was idealized and approximate. Our preceding discussion shows that he was aware of both the nature and size of the approximations and idealizations he was making. I am not, of course, claiming that Newton was a Bayesian. I do, however, believe that Bayesian analysis is justified and that it also explicates Newton's methods.

In this section I have tried to give a brief introduction to the Bayesian approach to confirmation of theories. I have tried to show that it can explicate scientific methods and that it is a fruitful and useful way of looking at science and its history. I have also attempted to demonstrate that Bayesians can deal with the criticisms raised against this approach.

In the next section I shall show how a Bayesian approach can illuminate the question why different experiments provide more confirmation of a hypothesis than repetitions of the same experiment.

4.4 WHY DO SCIENTISTS PREFER TO VARY THEIR EXPERIMENTS?

A proof

In general, before any experimental result is regarded as established, the experimental procedure is repeatedly instantiated in a variety of conditions: The instrumentation, apparatus, or analysis procedure is changed, or new laboratory personnel are involved. If the variation is sufficiently marked, the experiment ceases to be a mere repetition and becomes a different experiment entirely. It has often been remarked that repeating the same experiment (with the same result) yields diminishing returns of information. This is easily enough explained, for after sufficiently many repetitions, in a variety of circumstances, our confidence that we are genuinely witnessing the phenomenon in question has grown sufficiently great that further tests become unreasonable.

A rather different question is why, when we have repeatedly performed a test on a hypothesis h and satisfied ourselves that the observed result e is a genuine experimental result, where h, together with some background theory T, predicts e, we should perform, where possible, an entirely different type of test on h, involving, say, an entirely different range of background theory T'. Suppose we do, and obtain a result e', where $h \wedge T' \vdash e'$. We then regard e' as providing more confirmation of h than would be provided by merely another repetition yielding e. Can we explain this? Yes, and very simply, using a Bayesian analysis of the situation. Let us suppose that we have a conditional probability function relativized to a stock of background information including T and T'. Suppose E and E' are two different experimental procedures (in the sense to be discussed later) that can test h, each capable (in principle) of being indefinitely instantiated and yielding, respectively, $e_1, e_2, \ldots, e_n, e'_1, e'_2, \ldots, e'_m, \ldots$, where $P(e_i \mid h) = P(e'_j \mid h) = 1$ $(i, j \geq 1)$. The fact that E and E' are different experimental tests of h would seem to indicate that the results of using E' are less than maximally correlated with those of E, and vice versa. We can make this precise by defining E and E' to be different experimental procedures if and only if for all $m > m_0$, for some m_0,

$$P(e_{m+1} \mid e_1 \wedge e_2 \wedge \ldots \wedge e_m) > P(e'_{m+1} \mid e_1 \wedge e_2 \wedge \ldots \wedge e_m)$$

and for all $n > n_0$, for some n_0,

$$P(e'_{n+1} \mid e'_1 \wedge e'_2 \wedge \ldots \wedge e'_n) > P(e_{n+1} \mid e'_1 \wedge e'_2 \wedge \ldots \wedge e'_n)$$

where P represents the belief structure of the "ideal" experimenter.

The result we have mentioned, namely, that h receives more support from a mixture of confirming results relative to E and E' than from E or E' alone, is now simple to prove. For

$$\frac{P(h \mid e_1 \wedge \ldots \wedge e_k \wedge e'_{k+1})}{P(h \mid e_1 \wedge \ldots \wedge e_k)} = \frac{P(e_1 \wedge \ldots \wedge e_k)}{P(e_1 \wedge \ldots \wedge e_k \wedge e_{k+1})}$$

$$= \frac{1}{P(e'_{k+1} \mid e_1 \wedge \ldots \wedge e_k)}$$

Similarly,

$$\frac{P(h \mid e_1 \wedge \ldots \wedge e_k \wedge e_{k+1})}{P(h \mid e_1 \wedge \ldots \wedge e_k)} = \frac{1}{P(e_{k+1} \mid e_1 \wedge \ldots \wedge e_k)}$$

Hence, h receives proportionally more support at the $(k + 1)$th trial from experiment E' than from simply repeating E if and only if $P(e'_{k+1} \mid e_1 \wedge \ldots \wedge e_k) < P(e_{k+1} \mid e_1 \wedge \ldots \wedge e_k)$. But this, as we have seen, is precisely the case for all k greater than some k_0, where E and E' are less than perfectly correlated experimental procedures. And being less than perfectly correlated (in their outcomes) is a good formal explication of what one means by saying that E and E' are different experimental tests of h.

A plausible justification

In the previous section we demonstrated the happy conclusion that two different experimental procedures provide more evidence in support of a hypothesis than two repetitions of the same experiment. In this section we shall discuss some of the conditions under which it is reasonable to make such probability assignments, or, in other words, when we would consider the results of using E and E' less than maximally correlated. The following section will provide evidence, taken from the history of twentieth-century physics, to show that such assignments are implicitly, and sometimes explicitly, made by practicing scientists under the conditions we describe.

In this discussion we shall assume that the experiments referred to test the same hypothesis and measure the same quantity. It is, of course, true that the same hypothesis can be tested by measuring two different quantities. For example, the hypothesis "CP is violated in K_L^0 decays" can be tested either by measuring the rate of $K_L^0 \rightarrow \pi^+ \pi^-$ decays or by measuring the asymmetry in the decays $K_L^0 \rightarrow e^{\pm} \pi^{\mp} \nu$ or $K^0 \rightarrow \mu^{\pm} \pi^{\mp} \nu$. This can be dealt with by further restricting the hypothesis to two separate hypotheses: "CP is violated in K_L^0 decay into pions" and "CP is violated in the decay $K_L^0 \to \mu^\pm \pi^\mp \nu$." Even without such a restriction, these experiments will be different in the sense to be discussed later. An interesting example of two such different experiments that used the same physical apparatus simultaneously will be discussed in the next section.

Why should two experimental procedures be considered different if the experimental apparatuses are different? That this is not question begging or the beginning of infinite regress will be made clear later. I begin with two idealized and extreme categories of "different" experimental apparatuses. The first category consists of two apparatuses, A_1 and A_2, whose proper operation is implied by a single theory, T. These apparatuses may differ in size, material, geometric arrangement, number of particular pieces of experimental equipment, analysis procedure, and so forth. The second category consists of two apparatuses, B_1 and B_2, that depend exclusively on two distinct theories, T and T', respectively. Such theories can be distinguished by examining the set of statements implied by each of them. An example of this is the bubble chamber, which depends on the theory of bubble formation in superheated liquids, and the spark chamber, which depends on the theory of electric discharges in ionized gases. In practice, two experimental apparatuses may consist of parts that depend on the same theory and parts that depend on different theories. Such hybrid cases can be dealt with by considering the parts separately, using the arguments to be given later. In the following discussion we shall assume that our conditional probability functions are relativized to a store of background knowledge that includes T and T', as well as what we might call supertheories, such as quantum mechanics and special relativity, that are believed to apply to all of nature. The discussion is clearer, however, if we restrict ourselves to the specific theories of the apparatus.

Let us first consider a single apparatus A_1, that depends on a single theory T and that is operated at two different times t_1 and t_2. Then, unless some explicit time asymmetry is presupposed, a uniformity-of-nature assumption is made that implies that $P(e_i \mid e_j) = 1$, where e_i and e_j are similar results obtained at different times i and j, using A_1, and thus that $P(e_i \wedge e_j) = P(e_i)$. This assumption is, in fact, made for the vast majority of experiments, for no experiment acquires all of its data instantaneously, and there seems to be no useful distinction to be made concerning the time interval between data-acquisition runs, or whether or not the apparatus was turned off between these runs. In practice, any well-designed experiment will include checks to assure proper and constant operation of the apparatus (see the discussion of experimental checks in Chapter 6). We now consider two apparatuses, A_1 and A_2, that depend on a single theory T and for which the uniformity-of-nature assumption is made. We assert here that $P(e \mid e') < 1$, where e and e' are outcomes of A_1 and A_2, respectively. This is plausible, because successful operation of A_1 will not, in general, guarantee successful operation of A_2.[44] Thus, $P(e \wedge e') < P(e)$ or $P(e')$. From our previous discussion it then follows trivially that A_1 and A_2 give rise to different experimental procedures and hence are more informative, if successful, than either alone and are better than two Heraclitean or identical repetitions of a single experiment. The argument for two apparatuses dependent, respectively, on two distinct theories, T and T', now goes through a fortiori.

The discussion suggests that only identical, or Heraclitean, repetitions of a single experiment can be considered the same. Experiments involving different theories usually are thought of as giving rise to stronger inequalities, $P(e \mid e') \ll 1$, than those involving the same theory. I have not attempted here to give any numerical estimates of these probabilities or to deal with the question of how much support is given to a hypothesis by a given experimental result. This is a complex problem involving not only the precision of the result but also how much confidence one has that the apparatus is actually measuring the quantity involved (cf. the Galilean telescope and Aristotelian natural philosophers). For further discussion, see Chapter 6.

In addition, the degree of difference between two experimental results may very well depend on the theoretical context at a given time. To illustrate this, let us consider a hypothetical, idealized account of experimental tests of the Newtonian addition-of-velocities law prior to 1905, when classical mechanics was the only theory, and after 1905, when Einstein's special theory of relativity emerged as a serious rival. Prior to 1905 there would have been no reason to suspect that tests of this law at speeds close to c, the speed of light, would have been very different from tests at low speeds, because classical mechanics made no such distinction. After 1905, when special relativity gave a velocity-addition law that differed significantly from the Newtonian law at speeds close to c, we would think that experiments at high and low speeds would be quite different. In fact, it seems quite plausible that had such a high-velocity test given results in agreement with classical mechanics, then classical mechanics would have been more strongly confirmed after 1905 than before.

Our model can accommodate this quite easily. Let $e_1, \ldots, e_n$ be experimental results at low speeds in agreement with classical mechanics, and let e'_{n+1}, be a high-speed result in agreement with classical mechanics. It seems clear that prior to 1905, $P(e'_{n+1} \mid e_1 \wedge \ldots \wedge e_n) \approx P(e_{n+1} \mid e_1 \wedge \ldots \wedge e_n)$, and thus such a high-velocity test provides little additional confirmation, as compared with a repetition at low speed.

Let us now consider the situation after 1905. The background knowledge (relative to which our probabilities are defined) has now loosened up a bit, because special relativity has now emerged as a serious rival to classical mechanics. Let CM stand for classical mechanics, and SR for special relativity. Further define $P^*(\phi) = P(\phi \mid e_1 \wedge \ldots \wedge e_n)$, and let $P^*(\text{CM}) = p$ and $P^*(\text{SR}) = q$, where p and q, are both considerably greater than zero.

$$P^*(e'_{n+1}) = P^*(e'_{n+1} \mid \text{CM})P^*(\text{CM}) + P^*(e'_{n+1} \mid \text{SR})P^*(\text{SR}) + \epsilon_1$$

where ϵ_1 is quite small. (This appears to be justified historically.) But $P^*(e'_{n+1} \mid \text{SR}) = 0$, because special relativity predicts a very different result at speeds close to c than does classical mechanics. Thus, $P^*(e'_{n+1}) \approx P^*(\text{CM}) = p$, because $P^*(e'_{n+1} \mid \text{CM}) = 1$; that is $P(e'_{n+1} \mid e_1 \wedge \ldots \wedge e_n) \approx p$.

Now let e_{n+1} be a result in agreement with Newtonian law at low speeds. At such speeds, both classical mechanics and special relativity predict the same results, within observational error. Thus, we have, where ϵ_2 is also small in comparison with p and q,

$$P^*(e_{n+1}) = P^*(e_{n+1} \mid \text{CM})P^*(\text{CM}) + P^*(e_{n+1} \mid \text{SR})P^*(\text{SR}) + \epsilon_2$$
$$\approx 1 \times P^*(\text{CM}) + 1 \times P^*(\text{SR})$$
$$\approx p + q$$

That is, $P(e'_{n+1} \mid e_1 \wedge \ldots \wedge e_n) < P(e_{n+1} \mid e_1 \wedge \ldots \wedge e_n)$. Thus, we see that our model indicates that the emergence of rivals to a hypothesis *h* may very well determine what is or is not merely a repeated test of *h*.[45] We note also that these experiments will differ in respect to experimental apparatus, as discussed earlier.

Some illustrations

In this section I shall present evidence, taken from the history of twentieth-century physics, to support the view that practicing physicists do, in fact, use the procedures for classifying experiments as the "same" or "different" outlined in the previous sections.

Perhaps the clearest examples of almost Heraclitean repetitions were the multigroup bubble-chamber collaborations common in the 1960s.[46] In those experiments, sets of bubble-chamber photographs were taken under presumably identical conditions. In order to analyze the data more quickly, portions of those sets were then given to several groups of physicists at different locations. The results were then combined in a single report. In these cases, track measurements were made on similar measuring machines, and the track reconstruction and event fitting would be done with standard computer programs (e.g., TVGP, three-view geometric program, and SQUAW, an event-fitting program). Although some subjective judgments were made (e.g., track identification by ionization estimates), these were regarded as insignificant differences. The data were combined at one location, and the paper was written. Copies of the paper were circulated to members of the groups for comment, and a final version was agreed on. That led to the somewhat odd result that sometimes papers were published whose coauthors had never met. The existence of such combined reports illustrates not only the assumption of uniformity of nature in time [e.g., $P(e_i \wedge e_j) = P(e_i) = P(e)$] but also perhaps the assumption of uniformity of physicists.

A second rather interesting example of experiments that differed only slightly is provided by the lunar laser ranging experiments.[47] In that case, identical sets of data were analyzed independently by two groups. The reasons for such analysis were stated clearly by the Committee on Gravitational Physics:

[This] leads to computer programs with upwards of 100,000 Fortran statements. Verification of the reliability of such enormous programs is practically impossible. It would appear that only two sets of software, each developed completely independently, can serve as an adequate check on reliability.[48]

Thus, independent analyses (or "different" experiments) are regarded as providing more evidential weight than a single analysis done twice. It is interesting to note, however, that prior to publication of the results, but before all checks had been made, there was a discrepancy between the two results:

A valuable cross check for model or software errors has been provided by an MIT-AFCRL analysis of the lunar-ranging data with separately developed software. Both analyses presently give null results. However, up until the time when the MIT-AFCRL reported their null result to us, our solutions erroneously indicated a 1m amplitude for the Nordtvedt term. The error was traced to our truncation of some apparently small relativistic terms in the equations of motion. We planned to add all remaining relativistic terms of order $1/c^2$ before publication of our results, but we had not expected that these additions would affect the Nordtvedt term significantly.[49]

Even if the planned check had not been made, the discrepancy between the two results would no doubt have uncovered the error.

A similar case was presented by Aronson and associates[50] in discussing the possible energy dependence of η_{+-}, the CP-violating parameter in $K_L^0 \rightarrow \pi^+ \pi^-$ decay. There the same data were analyzed in two different ways by the same group of physicists. That those two independent analyses (different experiments, in our sense) were regarded as providing more evidence in favor of the hypothesis of energy dependence than would merely rechecking one analysis is shown in their statement:

The results obtained by Methods A and B are in good qualitative agreement, the principal difference being that Method B gives slope parameters which are somewhat smaller in magnitude and also in statistical significance. *Notwithstanding the agreement between the results of these two different methods* [emphasis added], the possibility remains that the apparent energy dependence of the K^0–$\overline{K}^0$ parameters is due to some unknown systematic effect in the data.[51]

Two "different" experiments can even be performed using the same apparatus at the same time. Cases in point are the experiments testing CP violation in $K_L^0 \rightarrow \pi^+ \pi^-$ decay[52] or observing asymmetry in $K_L^0 \rightarrow \mu^\pm \pi^\mp \nu$ decay.[53] There the apparatus guaranteed that two charged particles from K_L^0 decay traversed a magnetic spectrometer. A muon filter, consisting of one meter of lead and two banks of scintillation counters, was located behind the spectrometer. Only muons could penetrate the lead and count in both counter banks. The information whether or not these banks had fired was recorded for each event. In looking for $K_L^0 \to \pi^+ \pi^-$, only those events in which the counter banks had not fired (those without a muon) were used, although further analysis was required to identify $K_L^0 \to \pi^+ \pi^-$. To look for asymmetry in $K_L^0 \to \mu^\pm \pi^\mp \nu$ decay, only events that contained a muon (i.e., the counter banks had fired) were used. These experiments could, of course, have been performed separately, but that would have wasted valuable beam time.

The foregoing examples refer to "different" experiments in the sense that different analysis procedures or information sets were used. We now turn to cases that illustrate our remarks concerning the theoretical basis of the experimental apparatus. This discussion will use evidence from the "Review of Particle Properties,"[54] which is generally recognized as the standard source of information on particle properties for the physics community. We see clearly the importance of the type of apparatus used in a measurement by the fact that that information is given for each measurement of a quantity. Table 4.1 gives a partial list of measurement techniques used (specific devices such as the DESY Pluto detector are omitted). These may depend on different theories of operation (i.e., counters, spark chambers, emulsions, and bubble chambers). Within these groups we see differences in recording techniques (i.e., optical spark chambers in which data are recorded on film, as compared with automatic or wire spark chambers in which information is recorded electronically) and in materials used (i.e., deuterium, freon, hydrogen, helium, heavy liquid, propane, or xenon bubble chambers).

Tables 4.2 through 4.4 provide examples of measurements for which the apparatus used, in this case scintillation counters, depends on a single theory, but in which increasing specificity and differentiation are noted. Table 4.2 gives measurements of the percentage difference between the mean lives of the π^+ and π^- mesons divided by their average. Although scintillation counters were used for all the measurements, the geometries, beam conditions, and so forth, were different. Table 4.3 shows measurements of the mean life of the charged π mesons. Again, counters are used, but, now, in the light of possible, although not observed,

Table 4.1. *Types of experimental apparatus*

ASPK	Automatic spark chambers
CC	Cloud chamber
CNTR	Counters
DBC	Deuterium bubble chamber
ELEC	Electronic combination
EMUL	Emulsions
FBC	Freon bubble chamber
HBC	Hydrogen bubble chamber
HEBC	Helium bubble chamber
HLBC	Heavy-liquid bubble chamber
HYBR	Hybrid: bubble chamber and electronics
MMS	Missing-mass spectrometer
OSPK	Optical spark chamber
PBC	Propane bubble chamber
PLAS	Plastic detector
SPEC	Spectrometer
SPRK	Spark chamber
STRC	Streamer chamber
WIRE	Wire chamber
XEBC	Xenon bubble chamber

Source: Particle Data Group (note 54, p. S61).

Table 4.2 *Mean-life difference* $[(\pi^+ - \pi^-)/average] \times 10^{-2}$

	Value	Error	Measurement technique
	0.23	0.40	CNTR
	0.4	0.7	CNTR
	−0.14	0.29	CNTR
	0.055	0.071	CNTR
Average	0.053	0.068	

Source: Particle Data Group (note 54, p. S68).

differences in lifetimes between the two charges, the charge of the meson beam used is shown. A further difference is shown in Table 4.4, which lists measurements of the ratio of muon to proton magnetic moments. These are all counter experiments, but now not only is the charge of the muon given, but the specific phe-

Table 4.3. *Charged-pion mean life (units 10^{-9} sec)*

Value	Error	Measurement technique	Charge
25.6	0.8	CNTR	
25.46	0.32	CNTR	+
26.02	0.04	CNTR	+
25.6	0.3	CNTR	
25.9	0.3	CNTR	
26.67	0.24	CNTR	
26.04	0.05	CNTR	+
26.02	0.02	CNTR	+ −
26.09	0.08	CNTR	+
Average 26.030	0.023		

Source: Particle Data Group (note 54, p. S69).

nomenon used in the measurement is also shown (i.e., spin resonance, phase of precession, hyperfine splitting).

Measurements using apparatuses that depend on different theories are illustrated in Table 4.5, giving values for the left-right asymmetry in $\eta \to \pi^+ \pi^- \pi^0$. The apparatuses used were bubble chambers, both hydrogen and deuterium, and spark chambers, optical and automatic.

An interesting illustration of two "different" experiments occurred in attempts to find right-handed currents in muon decay. In that case, the effect was looked for by using two different methods, by essentially the same group of physicists (one additional graduate student participated in the second analysis). One method used the end-point spectrum of positrons from muon decay, and the other used the muon spin rotation. The physical apparatus remained almost the same, except that instead of a longitudinal magnetic field to fix the muon spin direction as in the first method, a transverse magnetic field was used to precess the spin in the second. The theories of the apparatus were thus different. The data were actually taken with an hour-long run of spin-precessed data interleaved between two runs of spin-fixed data. No effect was seen in either case, and the experimenters remarked that "The good agreement between the present μSR result [$\xi P_\mu \delta/\rho$, the quantity of interest, is greater than 0.9955] and the previous end-point rate analysis result ($\xi P_\mu \delta/\rho > 0.9959$) despite differences in the major sources of error, reinforces our confidence in each of them."[55]

Table 4.4. *Muon-to-proton magnetic-moment ratio*

Value	Error	Measurement technique	Charge	Comments
3.1865	0.0022	CNTR	+	Spin resonance
3.1830	0.0011	CNTR	+	Precession strobe
3.176	0.013	CNTR	−	Precession strobe
3.1834	0.0002	CNTR	+	Precession phase
3.18336	0.00007	CNTR	+	Precession strobe
3.1808	0.0004	CNTR	−	Precession strobe
3.18338	0.00004	CNTR	+	Precession phase
3.183351	0.000016	CNTR		Hyperfine splitting
3.183314	0.000034	CNTR		Hyperfine splitting
3.183330	0.000044	CNTR	+	Precession phase
3.183347	0.000009	CNTR	+	Precession phase
3.183336	0.000013	CNTR		Hyperfine splitting
3.183349	0.000015	CNTR		Hyperfine splitting
3.183326	0.000013	CNTR		Hyperfine splitting
3.1833467	0.0000082	CNTR	+	Precession phase
3.1833299	0.0000025	CNTR		
3.1833403	0.0000044	CNTR	+	Hyperfine splitting
3.1833448	0.0000029	CNTR	+	Precession strobe
Average 3.1833371	0.0000039			

Source: Particle Data Group (note 54, p. S65).

Table 4.5. *Left–right asymmetry parameter for* $\eta \rightarrow \pi^+ \pi^- \pi^0$
$(\times\ 10^{-2})$

Number of events	Value	Error	Measurement technique
1,351	7.2	2.8	DBC
1,300	5.8	3.4	HBC
10,665	0.3	1.0	OSPK
705	−6.1	4.0	HBC
36,800	1.5	0.5	ASPK
10,709	0.3	1.1	OSPK
1,138	−1.4	3.0	HBC
349	3.2	5.4	DBC
220,000	−0.05	0.22	ASPK
165,000	0.28	0.26	OSPK
Average	0.12	0.17	

Source: Particle Data Group (note 54, p. S71).

It is, however, true that experimental apparatuses thought to be the same, or at least very similar, because a single theory was thought to govern their operation have turned out to differ in significant ways. A case in point involves the experiments on double scattering of electrons in the 1920s and thereafter (see Chapter 2). The electrons used in those experiments came from both β decay and thermionic sources, and physicists of the day believed that the type of source used made no essential difference.[56] Later work, however, showed that β-decay electrons were longitudinally polarized, whereas thermionic electrons were unpolarized, which resulted in significantly different experimental results. In that same episode, another difference, thought to be insignificant at the time, was later shown to be important. That involved the question whether the electron beam scattered from the front surface of the foil targets (a reflection experiment) or passed through the foils (a transmission experiment). At the time, almost all such experiments were of a reflection type that avoided multiple-scattering problems. Later work, in which only the reflection and transmission characteristics of the experiment were changed, showed a clear difference in the results. In fact, the reflection type of apparatus masked an important experimental effect (expected on the basis of a theoretical calculation) that was observed only in transmission experiments after the significance

Table 4.6. η_{+-} *CP-violating*
parameter ($\times 10^{-3}$)

Value	Error	Measurement technique
1.95	0.20	OSPK
1.99	0.16	OSPK
1.92	0.13	OSPK
1.95	0.04	OSPK
2.00	0.09	OSPK
1.94	0.08	OSPK
Average 1.95	0.03	
2.23	0.05	ASPK
2.30	0.035	ASPK
2.25	0.05	SPEC
2.27	0.12	ASPK
Average 2.273	0.022	

Source: Particle Data Group (note 54, p. S90).

of the difference was realized. The fact that scientists, being fallible, are occasionally wrong in their judgments about significant differences between experiments does not argue against our model presented earlier. It shows only that differences may be difficult to find.

An interesting illustration of "different" experiments is shown in the current anomaly concerning the value of η_{+-}, the CP-violating parameter in $K_L^0 \rightarrow \pi^+ \pi^-$ decay. Prior to 1973, η_{+-} had been measured six times. The results were in good statistical agreement and had a mean of $(1.95 \pm 0.03) \times 10^{-3}$ (Table 4.6). Since 1973, there have been four additional measurements that have a mean of $(2.27 \pm 0.022) \times 10^{-3}$. As shown in Table 4.6, the first set of measurements all used optical spark chambers, whereas the second set all used automatic spark chambers (the spectrometer listed included such chambers). These sets are different in the sense discussed earlier, a point noted in the 1976 "Review of Particle Properties": "The newer experiments are *in principle superior* [emphasis added] (higher statistics, better acceptance, easier trigger conditions). The large discrepancy between the two sets of measurements is still unresolved and unexplained."[57] This will be discussed in further detail in Chapter 8.

The foregoing examples indicate that the physics community seems to follow procedures in agreement with my categorization of "same" and "different" experiments. Although I do not wish to argue that all institutionalized behavior of scientists has methodological justification, I do believe that this evidence reinforces my view. The examples have been taken from the "Review of Particle Properties," the standard reference for physicists, as well as from leading research journals. If the accepted "good" practice of science does not provide clues as to what "good" science should be, then what will? After all, it is generally agreed that science progresses. It seems unlikely that a random walk of good and bad methodological decisions would lead to a science that would progress, at least most of the time. I have also presented reasonable, independent justification for the model. In addition, this view leads quite simply to the conclusion, already intuitively appealing, that two different experiments provide more support for a hypothesis than do two repetitions of the same experiment. An application of this result will appear in Chapter 6 when we consider how we validate experimental results.

5

Do experiments tell us about the world?

If, as I have argued, experiment plays such an important role in physics, then we must have good reasons if we are to believe that experiments give us accurate and reliable information about the physical world. In the next four chapters we shall discuss several questions relating to this issue, including how we come to believe rationally in experimental results, and the question of fraud in science.

In this chapter I shall address a question raised by three historians of science, Peter Galison, Andrew Pickering, and Gerald Holton: the question whether or not the theoretical presuppositions of an experimenter influence the experimental results. There is at least an implicit suggestion that presuppositions may cause an experimenter to exclude data, to overlook unexpected results, and to either overlook or misestimate important sources of error. In Chapter 8 we shall examine the related issue of theoretical or experimental bandwagon effects: whether or not experimental results in disagreement with an accepted theory tend to be suppressed or excluded, and whether or not there is a tendency for experimental results to be in agreement with previous measurements.

In his study of gyromagnetic experiments in the early twentieth century, Peter Galison[1] points out that theory often provides experimenters with quantitative predictions that enable them to find the effects sought or to separate an effect from background sources of error. Such a prediction may also influence the decision to stop looking for such sources of error, declare the experiment ended, and report the result, which may be that predicted by theory. He cites Martin Deutsch, an experimental nuclear physicist, who wrote that "It is of course the ambition of every experimenter performing this kind of experiment to make a discovery, to sail safely between the Scylla of intellectual prejudice which makes us reject evidence not readily integrated without preconceived notions, and the Charybdis of irrelevance which has swallowed many working days spent in pursuit of instrumental artifice."[2]

Galison's case study suggests that in these early gyromagnetic experiments, theoretical presuppositions led to results that were incorrect, or at least in disagreement with currently accepted results. These preconceptions resulted in experimenters underestimating sources of error, or stopping the search for such errors when the result agreed with the theoretical prediction. The importance of these experiments led to many repetitions of the measurements and to the conclusion that these early results were wrong. I shall return to this point concerning repetition when I consider the issue of scientific fraud in Chapter 8.

Another case, considered by both Galison and Pickering,[3] is that of the experimental discovery of weak neutral currents. Events now attributed to such currents were seen in early experiments but were thought to be due to neutron background. At the time there was no theoretical prediction of such currents. Later, after theory did predict their existence, such currents were reported. Pickering seems to suggest that the currents were observed only because they were predicted. This seems to me to be somewhat unjustified. Theory cannot, after all, create the events. This episode illustrates the difficulty of observing nature. These experiments were at the forefront of their field, and the effect was quite difficult to observe. In the absence of any theoretical prediction it may have been wrong, but certainly not unreasonable, to overlook these events, especially when the observations were in agreement with plausible estimates of background. It is to the credit of the experimenters that the events were seen, even after they were predicted. Frank Sciulli, a leading neutrino experimenter, provided a very plausible summary:

In retrospect, it is likely that events due to neutral currents had been seen as early as 1967. Data from the CERN heavy-liquid bubble chamber . . . showed a surprisingly large number of events with hadrons in the final state, but with no visible muon. The calculations of neutron and pion-initiated backgrounds were so uncertain as to render the observed number of events inconclusive. In 1967 there was little pressure to rectify these uncertainties. Five years later the theoretical climate had changed dramatically, so that there were persistent but cautious efforts to conclusively resolve whether such events were actually anomalous.[4]

As I shall discuss in the next chapter, theoretical explanation can help to provide reasons for belief in an experimental result.

One of the important examples of such a strategy is the recent discovery of the $W^{\pm}$ bosons, in which the Weinberg-Salam unified theory of weak and electromagnetic interactions played an important role in establishing the experimental results. Even in such a theory-dominated experiment, the experimenters reported a completely new type of event that was not predicted by any existing theory: "We report the observation of five events in which a missing transverse energy larger than 40 GeV is associated with a narrow hadronic jet and of two similar events with a neutral electromagnetic cluster (either one or more closely spaced photons). We cannot find an explanation for such events in terms of backgrounds or within the expectations of the Standard Model."[5]

A somewhat different problem is the suggestion that theoretical preconceptions may cause an experimenter to exclude data that are in disagreement with those preconceptions. Gerald Holton seemed to suggest that that was the case in his discussion of Millikan's oil drop experiments.[6] He noted that although Millikan claimed to have published all of his data, an examination of Millikan's laboratory notebooks showed that that was not true. There is a hint that Millikan used his presupposition concerning the quantization of charge to select his data, and then used his data to support his presupposition. In the next section I shall present a detailed reexamination of Millikan's experiment to see if that was, in fact, the case.

5.1 MILLIKAN'S PUBLISHED AND UNPUBLISHED DATA ON OIL DROPS

Millikan's oil-drop experiments are justly regarded as a major contribution to twentieth-century physics. They established the quantization of electric charge and also measured that charge precisely.[7] Nevertheless, questions remain concerning Millikan's data and their analysis.

In presenting his final results in 1913, Millikan stated that the 58 drops under discussion had provided his entire set of data.

"It is to be remarked, too, that this is not a selected group of drops
but represents all of the drops experimented upon during 60 consecutive days*, during which time the apparatus was taken down
several times and set up anew."[8] Gerald Holton examined Millikan's laboratory notebooks for that period and noted approximately 140 separate runs. (My own count is 175 drops.) Even if
one is willing to count only observations made after 13 February
1912, the date of the first observation Millikan published, there
are 49 excluded drops. We are left with the disquieting notion
that Millikan selectively analyzed his data to support his preconceptions.

The calculations

The equation of motion for an oil drop falling under an upward
electric field F is $m\ddot{x} = mg - K\dot{x} - e_n F$, where e_n is the drop's
charge, and m is its mass compensated for the buoyant force of
air. According to Stokes's Law, which holds for a continuous
retarding medium, $K = 6\pi a\mu$, where a is the drop's radius, and
μ is the air's viscosity; to take into account the particulate character of air, Millikan replaced K by $K/(1 + b/pa)$, where p is the
air pressure, and b is a parameter fixed by experiment.[9] Because
all the measurements were made at terminal velocities, $\ddot{x} = 0$,
whence

$$e_n = \frac{mg}{F}\left(\frac{v_f + v_g}{v_g}\right) \tag{5.1}$$

Here the subscripts indicate terminal velocities without (v_g) and
with (v_f) the field, respectively. Now m can be replaced by a,
using $m = (4/3)\pi a^3(\sigma - \rho)$, σ and ρ being the densities of oil
and air, respectively; a can be done away with in favor of μ, using
Stokes's Law,[10] and the ratios of distance, d, to times of fall and
rise, t_g and t_f, can be substituted for the velocities. Altogether,

$$e_n = ne$$
$$= \frac{9\pi \, d\sqrt{2/g}}{F}\left[\frac{\mu^3}{(\sigma - \rho)(1 + b/pa)^3}\right]^{1/2}\left[\sqrt{v_g}\left(\frac{1}{t_f} + \frac{1}{t_g}\right)\right] \tag{5.2}$$

whence

$$e = \frac{A \cdot B(T) \cdot C(n, T)}{F} \tag{5.3}$$

B is a function of temperature T, because σ, μ, and ρ all depend on it. $C(n, t) = \sqrt{v_g}\,(1/n)(t_f^{-1} + t_g^{-1})$, which we write, for convenience, $C(n, t) = \sqrt{v_g}\,C'(n, t)$. The first equality in equation (5.2) expresses the assumption that charge is quantized.

An alternative calculation makes use of successive rise times, t_f, t'_f, in cases in which, in the intervening fall, the drop acquires an increment of charge Δn. In this case,

$$e = \frac{A \cdot B(T) \cdot D(n, t)}{F} \tag{5.4}$$

where

$$D = (\sqrt{v_g}/\Delta n)(t'_f{}^{-1} - t_f^{-1}) = \sqrt{v_g}\,D'(\Delta n, t)$$

We see that $D'(\Delta n, t)$ should equal $C'(n, t)$. That equality provided Millikan with a useful method of estimating $C'(n, t)$, because Δn is usually a small number.[11] At first he used only C', but by 10 February 1912, just before he started accepting data for publication, he began using both methods.

The data used in the recalculation of Millikan's work are contained in a microfilm of his laboratory notebooks obtained from the Millikan Collection at the California Institute of Technology.[12] The notebooks cover the period 28 October 1911 to 16 April 1912, ending with several undated pages that include some of Millikan's final calculations of e.

The notebooks contain observations on 175 oil drops, along with voltage and chronoscope corrections and measurements of the density of clock oil. Samples of the data sheets are shown in Figures 5.1 and 5.2. The columns labeled G and F show measurements of t_g and t_f, respectively. The average value of t_g and its inverse are given at the bottom of column G. To the right of column F are calculations of t_f^{-1} and of the difference $D'(\Delta n, t)$; farther to the right is $C'(n, t)$. The top of the page gives the date, the number and the time of the observation, the temperature θ, the pressure p, the voltage readings (including the actual reading plus a correction), and the time at which the voltage was read. On many pages, such as in Figure 5.2, Millikan goes on to calculate a and e. The data given earlier, combined with the physical dimensions of the apparatus, the densities of clock oil and of air,

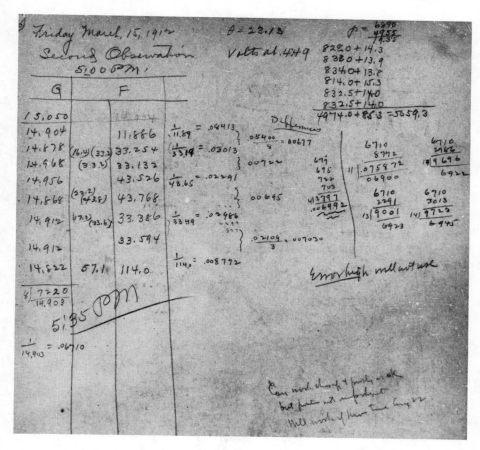

Figure 5.1. Millikan's data sheet for 15 March 1912 (second observation). Courtesy California Institute of Technology Archives.

the viscosity of air, and the value of g, are all that is required to calculate e.

In factor A [equation 5.3], the only physical quantities required are the constants g and d. For the former, Millikan used $g = 980.3$ cm/sec^2,[13] which agrees well with modern values calculated for Chicago; the latter he measured to one part in a thousand.[14] In order to determine F, the second factor in equation (5.3), we need both the voltage difference and the distance between the condenser plates. Millikan measured that distance, d_p, as 16.00 ± 0.01mm, and he used three different sets of corrections to the

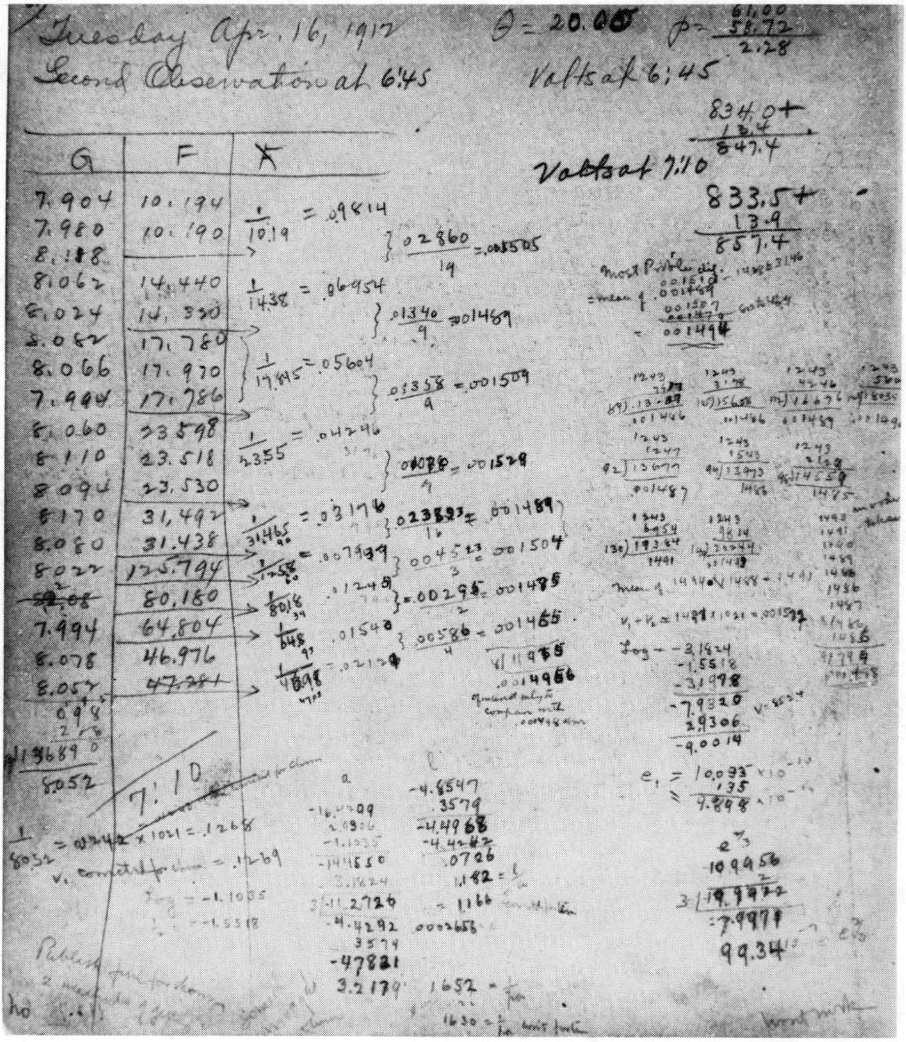

Figure 5.2. Millikan's data sheet for 16 April 1912 (second observation). Third anomalous event. Notice "Won't work" in lower right-hand corner. Courtesy California Institute of Technology Archives.

voltmeter readings.[15] I have used the set Millikan used in his final reworking of the data in August 1912 for all my recalculations. The differences between sets of corrections are quite small, about 0.05 percent. Although Millikan claimed that he had used the average of the voltage readings before and after the observations, he did not do so. He apparently exploited his knowledge of the experimental situation and adjusted the voltage used. I have consistently used the average value. This leads to differences between our calculations of no more than 0.1 percent.

To evaluate $B(T)$ in equation (5.3), we need both pressure and temperature measurements – the pressure to make the Stokes's Law correction, and the temperature to correct p, μ, and σ. On 28 October 1911, Millikan noted:

A correction of .0041% per degree C must be applied for change in viscosity of air, 23°C is the temperature where no correction is needed. Below 23°C, the correction must be subtracted. A correction of .0004% per degree must be applied for change in density of oil. This correction can be applied with 23°C as the point of no correction. It is also minus when the temperature is below 23°C. Correction for (change of viscosity + change of density) = .0045% [a slip for 0.45%] per degree C.

This correction is to be applied to the calculated value of e, not to the values of density or viscosity. Millikan measured the density of the oil twice, obtaining $\sigma = 0.9199$ g/cm^3 each time.[16] For μ, the viscosity of air at 23°C, he used 0.0001825.[17] Similarly, the value of the parameter b is given as "$b = .0006254$, p being measured for the purposes . . . of this calculation in mm. [a slip for cm.] of Hg at 23°C, and a being measured in cm."

The only correction to be applied to $C(n, t)$ arises from chronoscope errors. Millikan did not apply this correction consistently. The later observations were corrected, but the earlier data, including some of the published events, were not. I have applied the chronoscope correction only to t_g, which gives, in general, the larger part of $C(n, t)$. The error introduced by this limited correction is less than 0.1 percent.

The expected discrepancy between my calculations and those of Millikan is 0.15 percent. It arises from adding in quadrature the difference in voltage corrections, 0.05 percent, the difference in average voltage, 0.1 percent, and the discrepancy from limiting the chronoscope correction to t_g, 0.1 percent. The expected dis-

Table 5.1. *Comparison of Millikan's values (RAM) and Franklin's values (ADF) for* e

Drops	e (RAM) (× 10^{10} esu)	e (ADF) (× 10^{10} esu)	σ (RAM)[a]	σ (ADF)
Published				
First 23[b]	4.778	4.773	±0.002	±0.004
All 58	4.780	4.777	±0.002	±0.003
First 23[c]	4.776	4.773	±0.003	±0.004
All 58[c]	4.778	4.777	±0.002	±0.003
First 23[d]	4.784	4.782	±0.005	±0.006
All 58[d]	4.782	4.781	±0.003	±0.003
Unpublished				
To 2/13/1912		4.750		±0.010
After 2/13/1912		4.789		±0.007
Almost all drops[e]	4.781	4.780	±0.003	±0.003

[a]Statistical error in the mean.
[b]Events used by Millikan to determine e.
[c]With problematic drops corrected as in Table 5.2 (p. 154).
[d]With problematic drops corrected and with the excluded drops.
[e]The published 58, corrected, plus 25 unpublished measured after 13 February 1912.

crepancy is in good agreement with the actual discrepancy of 0.18 percent between Millikan's values for *e* and my own.

In order to get an accurate estimate of Δn, and thus n, an event must have several measured values of t_f, usually at least three or four. Otherwise one is forced to assume the value of *e* that one wishes to calculate. In several cases to be discussed later, both Millikan and I excluded events with too few values of t_f. My recalculations and Millikan's results are given in Table 5.1.

Selective analysis of data

The problem of measuring the charge on the electron consisted of two closely related parts: the first, determining whether or not electric charge is quantized; the second, accurately measuring the value of the charge. By 1913, Millikan regarded the question of charge quantization as settled on the basis of his own previously published work.[18] During the course of those measurements, he fought with Felix Ehrenhaft over the quantization of charge. In

1912 and 1913, a lull occurred in the controversy, and opinion favored Millikan. A younger colleague of Ehrenhaft's, Karl Przibram, wrote Millikan in 1912 acknowledging his earlier mistakes and stating his agreement with Millikan's work. Despite Przibram's concession, Ehrenhaft and two of his pupils, F. Zerner and D. Konstantinowsky, returned to the attack in 1914 and 1915.[19] Millikan answered the new criticism, which he regarded as essentially the same as Ehrenhaft's earlier criticism, by restating the evidence we are about to examine.[20] In this 1913 experiment, Millikan was concerned primarily with precise measurement of e.

In our examination we must also keep in mind that Millikan had far more data than he needed to decrease the error in the measured value of e by an order of magnitude. He gave his statistical error as follows: "The largest departure from the mean value found anywhere in the table amounts to 0.5 per cent., and the 'probable error' of the final mean value computed in the usual way is 16 in 61,000."[21] His quoted error in e of ± 0.009, for a value of $e = 4.774 \cdot 10^{-10}$ esu, is made up of systematic or experimental effects: the uncertainty in the viscosity of air, the uncertainty in the distance between the plates, the uncertainty in the voltmeter readings, and the uncertainty in the crosshair distance. He was correct to conclude that, in 1913, there was "no determination of e or N by any other method which does not involve an uncertainty at least 15 times as great as that represented in the above measurements."[22] Millikan did not even require all of his 58 published drops in computing his final value of e. He used only those drops (23 of them) for which the correction to Stokes's Law was smaller than 6 percent, to guard against any effect of an error in b.

Millikan had a good idea of the results he expected, although his expectations seem to have changed as he went along. He often made comments comparing his observed values with those he expected, such as "3% low." For the early events, these statements make sense only if he was comparing his observations to the value of $e = 4.891 \cdot 10^{-10}$ esu given in his paper of 1911. He also had available his published graph of $e^{2/3}$ against $1/pa$, a straight line, which enabled him to compare his observed values of e with those expected for the same values of $1/pa$. At some point Millikan realized that his new observations of e were consistently lower than his earlier value. In his first observation of 7 March 1912, he noted that "This is OK but volts are a little uncertain and tem[perature] also is bad. It comes close to lower line." Against the third observation of 16 March 1912, Millikan remarked, "Too high by 1½%." Because my calculated value of e for this event is $4.829 \cdot 10^{-10}$ esu, it would seem that he referred to a lower expected value for e.

Exclusion of events

In experiments before 13 February 1912, Millikan was concerned with getting his apparatus to work properly. He worried about convection currents inside his apparatus that could change the path of the oil drop. He made several tests on slow drops, for which the convection effects would be most apparent.[23] Millikan's comments on these tests are quite illuminating. On 19 December 1911, he remarked that "This work on a very slow drop was done to see whether there were appreciable convection currents. The results indicate that there were. Must look more carefully henceforth to tem[perature] of room." On 20 December: "Conditions today were particularly good and results should be more than usually reliable. We kept tem very constant with fan, a precaution not heretofore taken in room 12 but found yesterday to be quite essential." On 9 February 1912, he disregarded his first drop because of uncertainty caused by convection; after the third drop, he wrote that "This is good for so little a one but on these very small ones I must avoid convection still better." No further convection tests are recorded.

The results of my recalculation of the events from this first period are shown in Figure 5.3. The values of e fluctuate between approximately 4.5 and $5.0 \cdot 10^{-10}$ esu, a spread of 10 percent, and stabilize between 4.7 and 4.8 in the later events. Millikan calculated many of these drops, deduced the value of e, and made comments such as "2% high." The average value of e obtained from these 53 events is $(4.750 \pm 0.010) \cdot 10^{-10}$ esu, which is more precise than any measurement by any other technique available at the time.

In these recalculations, I excluded 15 of the 68 events: 5 because of equipment malfunction for which Millikan gave specific reasons, 4 more because Millikan scratched them out before doing any calculations (I assume he had good reasons, which he did not

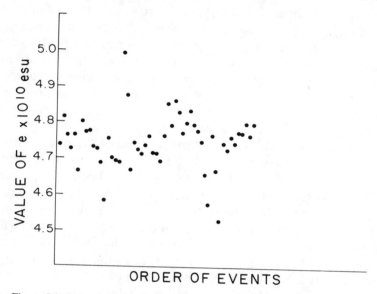

Figure 5.3. Values of *e*, events before 13 February 1912 (my calculations).

give, for distrusting these measurements), and 4 because they had only one or two time measurements taken with the field on.

Two excluded events call for more detailed explanation. The first of these is the third drop of 20 December 1911, one for which Millikan calculated a value of *e*. He got a very low value (my calculation gives $e = 4.325 \cdot 10^{-10}$ esu) and explained that the culprit "could not have been an oil drop." He probably believed that dust particles were responsible.[24] Although Millikan had taken precautions to exclude dust from his apparatus, it is not unlikely that one drop–and this was the only such drop he re-ported–might have been affected by dust. The second of these events, the fourth drop of 20 January 1912, is one for which I can find no consistent greatest common divisor of any reasonable size for either C' or D'. This drop will be discussed in detail later.

Millikan's use of the first drop of 13 February 1912 in his pub-lished results changes the situation. We must now assume that the apparatus was working properly, unless we are explicitly told otherwise. There are 107 drops in question, of which 58 were published. Millikan made no calculation of *e* on 22 of the 49 drops. I have excluded 6 from my recalculation for specific experimental reasons, such as wobbling of the drop, uncertainty in pressure, convection, and voltage irregularities. Three others were excluded because they had only one or two values of t_f. I have included 8 other events with few values of t_f but for which the estimate of n is straightforward, and 2 others that appear unproblematic. The most plausible explanation why Millikan did not calculate them is that when he performed his final calculations in August 1912, he did not need them for determination of e.

In two cases, Millikan performed some of the time calculations but did not proceed to obtain e. For the first of these, the third drop on 17 February 1912, he calculated both $C'(n, t)$ and D' ($\Delta n, t$) and remarked "Agreement poor. Will not work out." This is true only for D', where the values differ by about 5 percent. The values obtained for C' are quite consistent. Because Millikan had so much data, his decision to ignore this event was reasonable. The second case (Figure 5.1) concerns the second drop of 15 March 1912. Millikan again calculated both C' and D' and found that they differed by about 1 percent. That was too much: "Error high will not use. . . . Can work this up and prob[ab]ly is OK but point is not important. Will work if have time Aug. 22." Although there is a substantial difference between the results from the two methods of calculation, larger than any for which Millikan gave data in his published paper, it is no larger than the difference in some events that Millikan published, as shown in the laboratory notebooks. There is no reason to assume, as Holton seems to do, that there were unstated and perhaps unknown experimental reasons for exclusion of this event.

The 27 events that Millikan excluded and for which he calculated values of e are more worrisome. I have excluded 12 of these events from Figure 5.4 because their values of pressure and drop radius were such as to require a second-order correction to Stokes's Law. It is clear that Millikan suspected that the first-order correction to Stokes's Law did not suffice for these events.[25] Of the remaining 15 calculated events, Millikan excluded 2 on experimental grounds, 5 because they had three or fewer reliable measured values of t_f, and 2 for no apparent reason, probably because he did not need them for calculating e.

We are left with 6 drops. One is quite anomalous. In the other five cases, Millikan not only calculated values of e but also compared them with an expected value ("1% low"). For the four

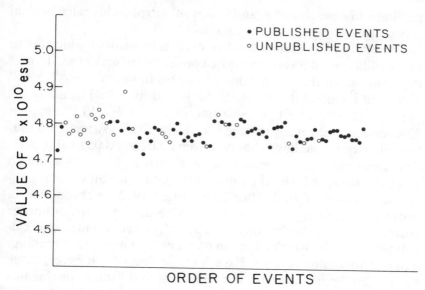

Figure 5.4. Values of e, events after 13 February 1912, both published and unpublished (my calculations).

earliest events, the comparisons make sense only if we use Millikan's value of 1911, $e = 4.891 \cdot 10^{-10}$ esu, whereas the last one ("Too high e by $1\frac{1}{4}\%$") seems to refer to a lower expected value. The four earliest events have values of pa that would place them in the group that Millikan used to determine e. His only evident reason for rejecting these five events is that their values did not agree with his expectations.

Millikan's analysis of one of these events, the fourth of 28 February 1912, does not show him at his best. He wrote that he had used D' "to correct" C' and that the procedure was "Justified in this case. Perhaps publish." But he provided no such justification and probably used the differences in rise times to improve the agreement between his measured value and his expected value of the fall time. Millikan's value of C' is given as 0.004032, whereas his value for D' obtained from the differences is 0.004000. In his first calculation he used the mean value, 0.004016, to obtain e. He later recalculated that event, taking 0.004000 as correct. My recalculation gives $e = 4.889 \cdot 10^{-10}$ esu, using a corrected t_g and $C' = 0.004042$. Millikan's value of 0.004000 for the differences gives $e = 4.838 \cdot 10^{-10}$ esu. Millikan was probably expecting a lower value of e, and that would explain his calculational procedure.

The effect of excluding the five events under discussion was to make Millikan's data appear more consistent, to make the "largest departure from the mean value anywhere in the table . . . 0.5%." If he had included these events, he would have had to allow a 2 percent departure. His mean value of $e = 4.778 \cdot 10^{-10}$ esu, for 23 events, would have changed to $4.784 \cdot 10^{-10}$ esu, and his statistical error would have changed from 0.002 to 0.005 (Table 5.1). The small effect of his selectivity is shown in Figure 5.4, which gives the values obtained by my recalculation for all events, published and unpublished, after 13 February 1912, in chronological order. It excludes the twelve events requiring a second-order correction to Stokes's Law, nine uncalculated events, one calculated event for which Millikan could not get a pressure reading, and two anomalous events. Except for one event that gave a value for e some 2 percent above the mean, there is little to distinguish the published events from the unpublished ones. My recalculation of e gives a mean value of $4.777 \cdot 10^{-10}$ esu for the 58 published drops, with a statistical error of 0.003; for the 25 unpublished drops included in the figure, the mean value of e is $4.789 \cdot 10^{-10}$ esu ($\pm$ 0.007). The number of unpublished events declined as Millikan went along, indicating the increasing reliability of his apparatus and his technique.

Selective calculations

Millikan's cosmetic surgery touched 30 of the 58 published events; for each of those 30, he excluded one or more (usually less than three) observations. For example, in the case of drop number 15, Millikan used only eight of the twelve measurements of t_f in calculating e.[26] My recalculation, using all of the data, gives a result little different from his. The same is true for all of these excluded observations. Millikan did not selectively exclude them on the basis of any discrepancy in his calculations, because, in general, he did not perform any calculations using them.

More serious, however, is Millikan's selective analysis of the data. He claimed that "in practically all of the following work in view of the large number of observations in the t_g column, the mean at the bottom of the column $[C'(n, t)]$ is considered more

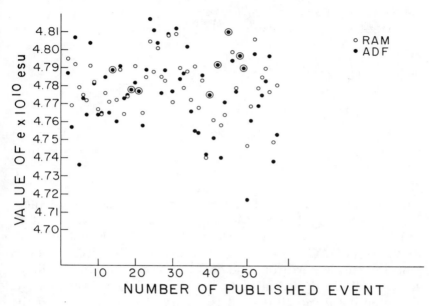

Figure 5.5. Comparison of Millikan's values and Franklin's values for e, 58 published events.

trustworthy than the bottom of the column $t_f [D'(\Delta n, t)]$ and it has been in fact exclusively used in the computation of e_1[our e]."[27] For at least 19 of the 58 published events, however, he used the average of C' and D', or some combination of the two that is not a strict average, or D' itself. In general, these effects are small, and the result of his tinkering was to reduce the statistical error rather than to change the mean value of e (Figure 5.5). For the first 23 drops, Millikan gets $e = (4.778 \pm 0.002) \cdot 10^{-10}$ esu, and I get $(4.773 \pm 0.004) \cdot 10^{-10}$ esu; for all 58, we get $(4.780 \pm 0.002) \cdot 10^{-10}$ esu and $(4.777 \pm 0.003) \cdot 10^{-10}$ esu, respectively (Table 5.1). Not only do the averages agree to about 1 part in 1,000, but the root-mean-square difference between Millikan's values and mine for 53 of the 58 published events is 0.18 percent, in good agreement with my prior appraisal of this discrepancy at 0.15 percent. The small remaining difference of 0.03 percent can be explained by slight differences in data and in calculational procedures. There are, however, rather larger discrepancies, about 1 percent, between Millikan's values and my own for pub-

Table 5.2. *Corrections to Millikan's events*

Drop	e (ADF) ($\times 10^{10}$ esu)	e(RAM) ($\times 10^{10}$ esu)	e (RAM), Corrected ($\times 10^{10}$ esu)	Discrepancy (%)
1	4.726	4.781	4.754	0.59
5	4.736	4.779	4.746	0.21
36	4.755	4.791	4.749	0.13
50	4.717	4.747	4.709	0.17
58	4.753	4.781	4.767	0.29

lished events 1, 5, 36, 50, and 58 (Table 5.2). In four of these events, excluding drop 1, Millikan used either the mean of C' and D' or D' alone. Table 5.2 shows the effect of using Millikan's stated formula, C', in each of these events. The discrepancy in drop 1 comes in part from one of the two serious errors Millikan made in all of these calculations: He used $\sqrt{v_g}$ instead of v_g in computing a (notebook, 13 February 1912). The rest of the discrepancy is due to a difference in the chronoscope correction to t_g.

The remaining discrepancies (Table 5.2) are in reasonable agreement within the limits set by the differences between my calculations and those of Millikan when one considers that the events were selected for their large discrepancies. The upshot of correcting Millikan's values of e for these events is to reduce his average value of e to $4.776 \cdot 10^{-10}$ esu for the first 23 drops, and to $4.778 \cdot 10^{-10}$ esu for all 58 drops, in comparison with my values of $4.773 \cdot 10^{-10}$ esu and $4.778 \cdot 10^{-10}$ esu, respectively (Table 5.1).

One may wonder about the purpose of these selective procedures when Millikan's quoted error in his value of $e = (4.774 \pm 0.009) \cdot 10^{-10}$ esu is made up of systematic or experimental effects: uncertainty in the viscosity of air, uncertainty in the distance between the plates, uncertainty in the voltmeter readings, and uncertainty in crosshair distance. Changing the statistical error from 0.002 to 0.004 would change the total error from 0.009 to 0.010, which is certainly insignificant.

Three anomalous events

Three anomalous events have been excluded from our analysis. The first of these, the fourth drop of 20 January 1912, has not yielded a value for the greatest common divisor of either C' or D' that has any reasonable size and that gives integral values for Δn and n. If we use the smallest difference between values of D' (namely, 0.001225) as the greatest common divisor, we get no integral values for Δn or n and a change in the charge of the drop as large as 85 units, which is totally inconsistent with any other observations. Millikan got roughly consistent values for C' by excluding one value of t_f: 28.466. He also miscalculated $v_g = 1.022/18.62 = 0.06910$ when it should have been 0.05483. His values for t_g are very inconsistent, and the difference between his two measurements of the 7-sec time, 7.364 and 7.824, is considerably larger than his usual measurement error. The whole event was unreliable.

Millikan did no calculations on the fourth drop of 7 March 1912. The values of C' and D' are all quite consistent. Using the mean value of $C' = 0.0029575$, we get $e = 2.022 \cdot 10^{-10}$ esu. I have consistently used $e = 4.78 \cdot 10^{-10}$ in calculating a, the drop radius. Using $e/3$ instead, I get $e = 1.915 \cdot 10^{-10}$ esu. This is not, however, evidence in favor of quarks or fractional charges. As the data show, favorable evidence would require not only that the charge on the drop be fractional, but also that the changes in the charge on the drop be fractional. It would indeed be remarkable if fractional changes in charge (there were no others, except for the third anomalous event, to be discussed later) occurred only on drops that were themselves fractionally charged. There are only two possible sources of error, σ and V, if one accepts the time measurements. It is very unlikely that an alteration in the effective value of σ owing to dust could have produced a 60 percent discrepancy. The other possibility is that the voltage was too large. Because Millikan used six banks of batteries on this drop, this explanation would require that four of the banks were somehow disconnected and miraculously reconnected before the next observations on 11 March. Millikan did not mention a failure or a miracle. I can find no plausible explanation for the anomalous result.

The same problem applies to the third anomalous event, the second drop of 16 April 1912 (Figure 5.2, Table 5.3). This event is the most worrisome of the three, because it is among Millikan's very best observations, as shown by the internal consistency of the values of C' and D' and by their agreement with each other. Millikan originally liked it: "Publish. Fine for showing two meth-

Table 5.3. *Recalculation of third anomalous event, 16 April 1912, second drop, $t_g = 8.052$ sec*

t_f	$\dfrac{1}{t_f}$	$\dfrac{1}{t'_f} - \dfrac{1}{t_f}$	Δn	$D'(\Delta n, t)$	$nC'(n, t)$	n	$C'(n,t)$
10.194	0.098116				0.222303	148	0.001502
10.190							
		0.028575	19	0.001504			
14,440	0.069541				0.193728	129	0.001502
14.320							
		0.013504	9	0.001500			
17.780	0.056037				0.180224	120	0.001502
17.970							
17.786							
		0.013572	9	0.001508			
23.598	0.042465				0.166652	111	0.001501
23.518							
23.530							
		0.010684	7	0.001526			
31.492	0.031781				0.155968	104	0.001500
31.438			16	0.001489			
		0.023821					
125.794	0.007950				0.132137	88	0.001502
		0.004522	3	0.001507			
80.180	0.012472				0.136659	91	0.001502
		0.002959	2	0.001480			
64.804	0.015431				0.139618	93	0.001501
		0.005856	4	0.001464			
46.976	0.021287				0.145474	97	0.001500
		Average		0.001497			0.001502

Note: a (using *e*) = 0.000265; *a* (using 2*e*/3) = 0.00231; *e* = 3.089 · 10^{-10} esu; *e* (second-order correction) = 2.8 · 10^{-10} esu.

ods of getting *v*." Using the calculated value of *C'*, I obtain *e* = 3.430·10^{-10} esu. Correcting *a* by using *e* = 2*e*/3, I get *e* = 3.089·10^{-10} esu. With a second-order correction to Stokes's Law, *e* = 2.810·10^{-10} esu. Once again, both the charge on the drop and the changes in charge must be fractional, a highly unlikely occurrence. Once again, neither dust nor voltage problems can explain the anomaly. Millikan remarked: "Something wrong w[ith] therm[ometer]," but there is no temperature effect that could by any stretch of the imagination explain a discrepancy of this magnitude. Millikan knew that the value of *e* was wrong, as shown by the comment "Won't work" at the bottom right of the data page (Figure 5.2).

There is an explanation for these anomalous events. Later work, to be discussed in the next section, showed that despite Millikan's claim that his method was good for total charges up to about 150*e*, the method starts to become unreliable for total drop charges of about 20*e* and is almost totally unreliable for total drop charges above 30*e*. The two anomalous drops in question, 7 March 1912 (fourth drop) and 16 April 1912 (second drop), had average charges of about 30*e* and 110*e*, respectively. The failure to get any reasonable result for C' or D' on the other excluded event also indicates that its total charge was large.

Discussion

There can be no doubt that Millikan excluded at least some of his data on the basis of his preconceptions. The five excluded events, discussed earlier, increased his statistical error, but did not change the average value of *e* significantly. Millikan probably knew this, at least qualitatively. Similarly for his selective calculational procedures. Such an excuse is not available for his exclusion of the third anomalous event. He probably excluded it to avoid giving ammunition to Ehrenhaft in the quantization controversy. Although, in retrospect, the exclusion of that event seems justified on the basis of its large total charge, Millikan did not know that. As I shall discuss in Chapter 8, a safeguard against fraud is repetition of theoretically important experiments. The same safeguard works here to guard against undue influence of theoretical preconceptions. This experiment has been repeated often, because quantization of charge is an important question, and also because knowledge of the precise value of the fundamental unit of charge is needed for precise calculations.

5.2 DID MILLIKAN OBSERVE FRACTIONAL CHARGES ON OIL DROPS?[28]

Although the work to be discussed later has, I believe, provided an explanation for Millikan's unpublished observation of fractional charge, a question still remains whether or not observations of fractional charge are contained in his data. This has become of more interest because of recent reports[29] of observation of fractional charges on superconducting niobium spheres. Although, as we shall see later, there is now some reason to question these results, it is still interesting to try to use the history of physics to cast light on a contemporary problem. In 1910, Millikan himself reported that "I have discarded one uncertain and unduplicated observation apparently on a singly charged drop, which gave a value of the charge on the drop some 30 per cent lower than the final value of e."[30] He explained that on the basis of rapid evaporation of the water droplet on which the observation was made.[31]

In this section I shall reanalyze Millikan's data to examine the evidence for both charge quantization and fractional residual charge on the oil drops. Charge quantization will be shown by looking at the changes in the charges on the oil drops, a point Millikan emphasized.[32] I looked for fractional residual charges by two methods: First, I examined the intercept generated by fitting a least-squares straight line to the data in n versus $(1/t_g + 1/t_f)$ for each drop (see Section 5.1 for details); second, I calculated, for each drop, the average deviation from integral charge obtained by dividing the total charge on the drop, for each individual measurement, by the best modern value for e, $4.80324 \cdot 10^{-10}$ esu.

The data used in this recalculation are contained in a microfilm of Millikan's laboratory notebooks obtained from the Millikan Collection at the California Institute of Technology.[33] The notebooks cover the period 28 October 1911 to 16 April 1912. In the recalculation, I have used only drops measured after 13 February 1912, the first drop published by Millikan. Prior to that date, Millikan had doubts about the proper operation of his apparatus and was, in particular, concerned about convection problems. Of the 107 drops in that period, I excluded 8 for the experimental reasons given by Millikan, such as drop flickering, irregular voltage, temperature changing rapidly, or inability to obtain a pressure or temperature reading.[34] Fifteen others were excluded because they had values of pressure and drop radius that required a second-order correction to Stokes's Law.[35] Regarding calculation of the change in charge on an oil drop, four other events were excluded because each had only one value of the total charge and thus could not give a value for the change in charge. A total of 80 events remained. I then calculated the value for ΔQ for

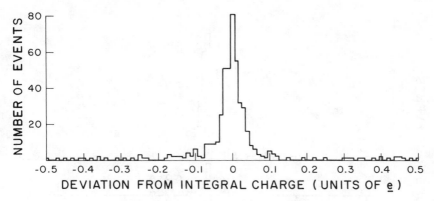

Figure 5.6. Deviation from integral charge computed from the change in charge, ΔQ, of the oil drop.

each pair of measurements of t_f and t'_f using equation (5.4). The results of the calculations are shown in Figure 5.6.

I examined the evidence for fractional residual charge on the individual oil drops by two methods. As can be seen from equation (5.3), n should be a linear function[36] of $(1/t_g + 1/t_f)$.[37] I fitted a least-squares straight line to n as a function of $(1/t_g + 1/t_f)$ of the form $n = a(1/t_g + 1/t_f) + b$. The intercept b at $(1/t_f + 1/t_g) = 0$ gives the fractional residual charge on the drop. For these calculations, in addition to the 27 drops excluded earlier, I further excluded 19 drops that did not have at least four unique values of n.[38] This was to ensure a reasonable error in the least-squares fit. A total of 61 drops remained. In addition, I excluded 13 individual measurements of t_f, out of a total of 804.[39] As seen from Figures 5.7 and 5.8, the determination of e becomes unreliable for values of $n > 20$. Seven of the excluded points were on drops with large charges and for which we could not uniquely assign a value of n. Six other points had $n < 20$, but have large deviations from integral charge (closer to half-integral n), an unlikely occurrence, and were considered bad observations.[40]

A second method of calculating residual charge was by calculating Q using equation (5.2) and dividing by the best modern value of e, $4.80324 \cdot 10^{-10}$ esu. I then calculated the average deviation from integral charge for each drop using the same 61 drops as before.

These two methods have different advantages. The least-

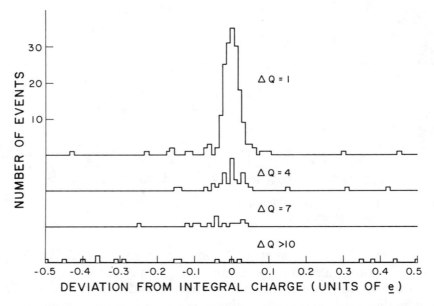

Figure 5.7. Deviation from integral charge for different values of the change in charge, ΔQ, on the drop.

squares fit is less sensitive to absolute errors in voltage, temperature, pressure, and viscosity of air, as well as to the Stokes's Law correction, than is the average-deviation method. The latter, however, is less sensitive to the effects of a single bad datum point, slow drifts in voltage, and evaporation.[41] Thus, both methods have been used.[42]

The results of deviations from integral values of charge obtained from calculation of ΔQ, the change in charge, are given in Figure 5.6. The distribution has a large peak at zero, as expected, and there is no evidence for fractional change in charge, which would be indicated by a peak at $\pm \frac{1}{3}$. There is also no evidence for any smaller unit of charge. As Millikan himself noted,[43] the changes in charge provide strong evidence for charge quantization. Figures 5.7 and 5.8 show the same results for various values of ΔQ, the change in charge, and $\overline{Q}$, the average charge on the drop. Both graphs show more deviation from integers as both the average charge and the change in charge increase, indicating the increasing unreliability of the method for drops with large total charges.

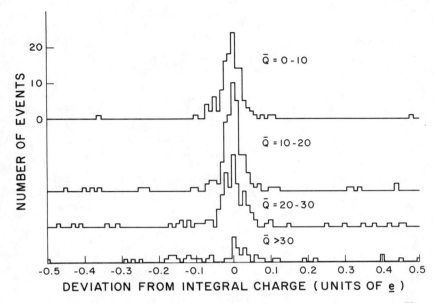

Figure 5.8. Deviation from integral charge for different average values $(\overline{Q})$ of the charge on the drop.

The results obtained for residual charges on the drops from both the least-squares fit and the average deviation are given in Figures 5.9 and 5.10, respectively. We have plotted only those values for $Q \leq 28$ because of the unreliability of these methods for large values of the total charge. There is also a larger error in the value for the residual charge as Q increases, as shown in the figures. No convincing evidence of fractional residual charge is seen for any of these drops. If we adopt as plausible criteria for fractional charge that both methods, least-squares fit (LSF) and average deviation (AD), give results closer to $\pm \frac{1}{3}e$ than to zero, and that both methods give results at least two S.D. from zero, we find two candidates. These are the events of 2 March 1912 (second observation) $(\overline{Q} = 21.1)$ and 29 March 1912 (second observation) $(\overline{Q} = 20.5)$. The values obtained for fractional residual charge for these events are $(-0.778 \pm 0.209)e$ (LSF), $(-0.272 \pm 0.070)e$ (AD), and $(-0.199 \pm 0.037)e$ (LSF), $(-0.186 \pm 0.016)e$ (AD), respectively. In addition, we find that the values for e obtained by using $(n - \frac{1}{3})$ rather than n give a

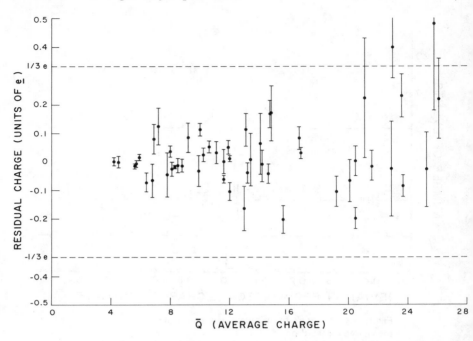

Figure 5.9. Residual charge on an oil drop as a function of the average charge $\overline{Q}$ on the drop computed by the least-squares method.

better fit to the modern value for e. For the 2 March event, we calculate $e = (4.818 \pm 0.015) \cdot 10^{-10}$ esu (using $n - \frac{1}{3}$) and $e = (4.742 \pm 0.020) \cdot 10^{-10}$ esu (n); for 29 March, $e = (4.838 \pm 0.007) \cdot 10^{-10}$ esu $(n - \frac{1}{3})$ and $e = (4.759 \pm 0.008) \cdot 10^{-10}$ esu (n). Although these two events are somewhat more consistent with an interpretation of fractional charge rather than with integral charge, the evidence is not strong. Both events also have reasonably large average charges, $\overline{Q} = 21.1$ and 20.5, respectively, where the calculational methods are becoming less reliable.

I conclude that Millikan's original data give strong evidence for charge quantization and no convincing evidence for fractional residual charge, although two events (out of sixty-one) are consistent with such an interpretation.

5.3 HAS ANYONE OBSERVED FRACTIONAL CHARGES?

In the last section we mentioned the attempts by Fairbank and his collaborators to demonstrate the existence of fractionally

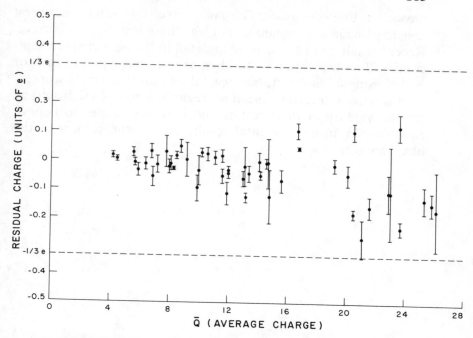

Figure 5.10. Residual charge on an oil drop as a function of the average charge $\bar{Q}$ on the drop computed from the average deviation of the total charge on the drop from integral values.

charged objects. This has been an important question since the formulation of the quark model of strongly interacting particles, or hadrons, by Gell-Mann and Zweig,[44] independently, in 1964. One consequence of this increasingly successful theory[45] was that the fundamental constituents of the hadrons, the quarks, would be fractionally charged, that is, having charge $\pm\frac{1}{3}e$ or $\pm\frac{2}{3}e$. Experimental searches for these quarks were conducted, and by the early 1970s a consensus was reached that they had not been observed, a consensus that has continued to the present.[46] That led, in that period, to the idea of quark confinement, that the force between quarks increased with distance, and that free quarks could never be observed.

The major anomaly for that theory has been the work of Fairbank cited earlier. That work has been in disagreement not only with an empirically successful and theoretically attractive theory but also with the results of other experimental searches.[47] Until recently there had been no resolution of the issue. Whereas most physicists, I believe, reject Fairbank's results, they have not been able to provide any explanation of why these results are incorrect. Recent results, to be discussed in detail in the next chapter, concerning the observation of residual charges not equal to zero or $\pm \frac{1}{3}e$, using a blind test, have cast doubt on Fairbank's work.

This raises interesting questions regarding how physicists argue for the validity of their results and how one comes to believe rationally in an experimental result. That is the subject of the next chapter.

6

The epistemology of experiment

Although all scientists and philosophers of science are agreed that science is based on observation and experiment, very little attention has been paid to the question of how we come to believe rationally in an experimental result,[1] or, in other words, to the problem of the epistemology of experiment.[2] How do we distinguish between a result obtained when an apparatus measures or observes a quantity and a result that is an artifact created by the apparatus? In this chapter I suggest that there are various strategies that both provide justification for rational belief in an experimental result and are used by practicing scientists. We shall be concerned here primarily, although not exclusively, with observations or results that are interpreted within an existing theory. Although a devout skeptic might doubt the validity of pointer readings or chart recordings, I do not believe that is the important question about the validity of experimental results.

Ian Hacking[3] has made an excellent start on discussing these issues. He points out that most modern experiments involve complex apparatus, and so, at the very least, the results are loaded with the theory of that apparatus. Dudley Shapere[4] has extended the idea of "direct observation" to include theoretical beliefs explicitly. In his discussion of the solar neutrino experiments, he stated that "X is directly observed if (1) information is received by an appropriate receptor and (2) that information is transmitted directly, i.e. without interference, to the receptor from the entity X (which is the source of the information)."[5] The dependence on theory is clear. Theory tells us what an appropriate receptor is and that the information is transmitted without interference. This issue of theory-ladenness of observation was discussed earlier when we considered the role of experiment in theory choice. I shall demonstrate later that for validating experimental results it is not a serious problem.

6.1 EPISTEMOLOGICAL STRATEGIES

Hacking suggests, in his discussion of the microscope, that if something can be observed using "different" microscopes (his example is dense bodies in cells), then it is real. He argues that it would be a preposterous coincidence if the same patterns were produced by two totally different kinds of physical systems (i.e., ordinary, polarizing, phase-contrast, fluorescence, interference, electron, or acoustic microscopes). "Different" here is clearly theory-laden. It is our theory of light and of microscopes that tells us that these are different. This does not, however, invalidate Hacking's argument. As discussed in Chapter 4, a hypothesis h receives more confirmation from "different" experiments than from repetitions of the "same" experiment.[6] Suppose that E and E' are two different experimental procedures, in the sense to be discussed later, that can test h, each capable (in principle) of being indefinitely instantiated and yielding, respectively, e_1, e_2, ..., e_n, ..., e_1', e_2', ..., e_m', ..., where $P(e_i \mid h) = P(e_j' \mid h) = 1$, $(i, j \geq 1)$. The fact that E and E' are "different" experimental tests of h would seem to indicate that the results of using E' are less than maximally correlated with those of using E, and vice versa. We can make this precise by defining E and E' to be distinct experimental procedures if and only if for all $m > m_0$, for some m_0,

$$P(e_{m+1} \mid e_1 \wedge e_2 \wedge \ldots \wedge e_m) > P(e_{m+1}' \mid e_1 \wedge e_2 \wedge \ldots \wedge e_m)$$

and for all $n > n_0$, for some n_0,

$$P(e_{n+1}' \mid e_1' \wedge e_2' \wedge \ldots \wedge e_n') > P(e_{n+1} \mid e_1' \wedge e_2' \wedge \ldots \wedge e_n')$$

where P represents the belief structure of the "ideal" experimenter.

The result we mentioned, that h receives more support from a mixture of confirming results relative to E and E' than from E or E' alone, is then easy to prove. We obtain

$$P(h \mid e_1 \wedge e_2 \wedge \ldots \wedge e_{k+1}') > P(h \mid e_1 \wedge e_2 \wedge \ldots \wedge e_{k+1})$$

if and only if

$$P(e'_{k+1} \mid e_1 \wedge e_2 \wedge \ldots \wedge e_k) < P(e_{k+1} \mid e_1 \wedge e_2 \wedge \ldots \wedge e_k)$$

Here the hypothesis will be of the form "the value of X is a," or "A has been observed." We also argued that at least one reason for making such probability assignments is that the theories of the apparatus are different.[7] If the theory is the same, then one can point to differences in size, in geometry, or even in the experimenters, and so forth. It seems clear, then, that here theory-ladenness[8] is a virtue rather than a defect.

A variant of the use of different experimental apparatuses is that of indirect validation. Suppose we have an observation that can be made using only one kind of apparatus. Let us also suppose that the apparatus can produce other similar observations that can be corroborated by different techniques. Agreement between these two different techniques gives confidence not only in the observations but also in the ability of the first apparatus to produce valid observations. This, then, provides an argument in support of the observation made only with that apparatus. An example of this is observation of the microtrabecular lattice using electron microscopy, which will be discussed in detail later. Scientists argue that other objects of similar size (e.g., microtubules) have been seen by both electron microscopy and with an ordinary light microscope. This should then support the idea that electron microscopy can detect objects of this size, and thus help to validate the observation of the lattice. Here the similarity of the observations, in this case in size, is of importance. The ability of the electron microscope to detect objects of very different sizes, particularly of larger size, would not be of much assistance in arguing for the lattice. In fact, it is the theory of the apparatus that indicates that size is an important parameter.

A more difficult and more interesting problem, discussed only briefly by Hacking, is how one deals with phenomena that can be observed with only one technique (i.e., electron microscopy or radioastronomy), or how one validates a single experiment. Hacking points out that the theory of the microscope changed drastically in 1873 when Ernst Abbe showed the importance of diffraction in its operation.[9] The use of microscope images remained robust despite such a theory change. Hacking attributes this to the fact that experimenters "intervened," that is, stained, injected fluid, and in other ways manipulated the objects under observation. This is, I believe, a special case of a more general strategy in which one predicts what will be observed after the intervention if the apparatus is working properly or is working as one expects it to. When the observation is made, we increase our belief both in the proper operation of the apparatus and in its results.[10]

This strategy is not available for radioastronomy and can be used only with difficulty for electron microscopy. Scientists can intervene by, for example, changing the temperature of the observed cells, or by inducing changes in the activity of the cell by chemical means, as will be discussed later. One may argue that when an electron microscope reveals the same things as an ordinary microscope, or a radiotelescope reveals the same things as an ordinary telescope, that gives us some confidence both in the observations and in the technique, and that is true. However, that does not seem totally satisfactory. It is precisely the phenomena visible only with an electron microscope or only with a radiotelescope that are of greatest interest. How, then, do we validate such observations? Extrapolation is notoriously dangerous. One strategy involves the use of a well-corroborated theory of the apparatus. If, as is indeed the case for both the electron microscope and the radiotelescope, the proper operation of the apparatus depends on such a theory, then the evidence supporting the theory also gives reason to believe the observations.

Even without such a theory, one can validate the observations. Consider the Galilean telescope. For centuries, Cremonini and other Aristotelians of the early seventeenth century have been ridiculed because they refused to look through Galileo's telescope to observe the moons of Jupiter. Even some of those who did denied their existence. Their skepticism was not without some merit. Galileo's telescope was not very good, and it was only his expertise in using it that enabled him to observe those moons. Even granting the observation of specks of light, how do we assert their real existence as moons of Jupiter, not as artifacts created by the telescope? One cannot here resort to the theory of the telescope and its independence of the distance to the object, because no such theory existed at the time of Galileo's *Starry Messenger*. One could argue that on earth, where one could check by direct observation, the telescope seemed to give valid images. But extrapolation to astronomical distances involves many orders of magnitude. I suggest here that it is the observed phenomena themselves that argue for their validity. Although one might imagine that a telescope could create specks of light, it hardly seems possible that it would create them in such a way that they would appear to be a small planetary system with eclipses and other consistent motions. It is even more preposterous to believe that they would satisfy Kepler's Third Law (R^3/T^2 = constant), although that argument would not have been available until publication of Kepler's *Harmonices Mundi* in 1619, and perhaps not until later in the seventeenth century, when it was generally accepted as a law.

A similar argument was offered by Robert Millikan to support his observation of the quantization of electric charge. He "found in every case the original charge on the drop an exact multiple of the smallest charge which we found that the drop caught from the air. The total number of changes which we have observed would be between one and two thousand and *in not one instance has there been any change which did not represent the advent upon the drop of one definite invariable quantity of electricity or a very small multiple of that quantity*" (emphasis in original).[11] The consistency of the data argued for their validity and against their interpretation as artifacts. No remotely plausible malfunction of the apparatus could produce such a consistent result. Millikan gave similar arguments in his other papers on electric charge.[12]

Interestingly, the consistency of the data provided support for Fairbank's recent reports of observations of fractional charge of $\frac{1}{3}e$: "Out of 26 repeat measurements, we have observed 11 residual-charge changes, in *every case* [emphasis added] of $\pm \frac{1}{3}e$."[13] The experimenters also stated that the residual charges on their spheres "fall into three groups which have weighted averages of $(-0.343 \pm 0.011)e$, and $(+0.001 \pm 0.003)e$, and $(+0.328 \pm 0.007)e$."[14] The fact that both total charges and changes in charge had only the values zero or $\pm \frac{1}{3}e$ supports their conclusion that fractional charges exist. Were other values to be observed, then one might doubt both their observations and their conclusions.

The fact that the consistent values of fractional charge observed were those predicted by theory added to their believability. Had these consistent values been $\pm \frac{1}{2}e$, a value not predicted by theory, one suspects that they would have been treated more skeptically.

It is ironic that recent work by Fairbank's group[15] gave results that were not zero or $\pm\frac{1}{3}e$. One problem in this experiment is that of possible experimenter bias. The results of the measurements of residual charge are known to the experimenter when the final data selection is made. To guard against possible bias, Luis Alvarez suggested that a random number unknown to the selector be added to each result and subtracted only after final-event selection was made. That was done, and the results of that blind test were $(+0.189 \pm 0.02)e$ and $(+0.253 \pm 0.02)e$. That procedure could not be used for all the niobium spheres, and a similar charge was observed in such an event. That observation of charges that were not zero or $\pm\frac{1}{3}e$ cast doubt on Fairbank's results. Professor Fairbank told me that after those discordant results were obtained from both the blind test and normal runs, the experiment was carefully reexamined. An instrumental effect was found that might account for the discrepancy. Tests are currently in progress. The question of the existence of fractional charges is still unresolved. The point here, however, is that the consistency of the data was used to argue that the results were not artifacts of the apparatus. It should be emphasized that these strategies support rational belief in a result. Rational belief may still be wrong.

Difficulties also attend validating observations with a radiotelescope. Certainly the radiotelescope can detect some, but not all, of the sources visible with an optical telescope. It is also based on a well-corroborated theory. I suggest that, in addition, part of the argument for validity of these observations is that they can be explained using the existing, accepted theory of the phenomena.

An interesting example of this, outside radioastronomy, is the recent discovery of the $W^{\pm}$,[16] the charged intermediate-vector boson, required by the independently well corroborated Weinberg–Salam unified theory of electromagnetic and weak interactions. Although the two experiments used very complex apparatuses and used other epistemological strategies, including independent checks on the proper operation of parts of the apparatus, independent methods of calculating the effect from the data, and eliminating possible sources of background that might simulate W decay, I argue that the theoretical explanation of the observations and the agreement with theoretical predictions helped, in part, to validate the observations.

The importance of the theory is shown even in the design of the particle accelerator used in the experiment. "The CERN Super Proton Synchrotron (SPS) Collider, in which proton and antiproton collisions at $\sqrt{s} = 540$ GeV (center of mass energy) provide a rich sample of quark-antiquark events, has been designed with the search [for $W^{\pm}$ bosons] as the primary goal."[17] The theoretical predictions are also clearly evident in the design of the experiments, the selection of events, the analysis of the data, and the validation of the observations. Both experiments were designed to detect electrons with high momentum transverse to the proton-antiproton directions, as predicted by theory: "We report here the results of a search for single electrons of high transverse momentum (P_T) which are expected to originate from reaction (1). [Reaction (1) was $\bar{p} + p \rightarrow W^{\pm}$ + anything followed by $W^{\pm} \rightarrow e^{\pm} + \nu\,(\bar{\nu})$.] Because the neutrino from W decay is not detected, the events from reaction (1) are expected to show a large missing transverse energy (of the order of the electron P_T) along a direction opposite in azimuth to that of the electron."[18] It should be emphasized that no other reaction known was expected to produce such high-P_T electrons. Events with high-P_T electrons, which had no jet structure, as required by theory, were selected as the final-event sample. These events also showed evidence of missing transverse energy of the right magnitude for W decay. From these events the mass of the W could be calculated, giving 81 ± 5 GeV/c^2 and 80^{+10}_{-6} GeV/c^2 for the UA1 and UA2 experiments, respectively. The Weinberg–Salam theory predicted the mass of the W to be 82 ± 2.4 GeV/c^2. The experimenters noted that their measurements were "in excellent agreement with the expectation of the Weinberg–Salam model."[19] In addition, both the number of observed events and the transverse-momentum distribution of the W's were in good agreement with theoretical predictions. It seems clear that the observations confirmed the theory. It also seems clear, as confirmed by discussions with colleagues working in high-energy physics and with a member of the experimental group, that the agreement with theory helped to validate the observations, which included examples of the other strategies suggested in this chapter. I have not done justice here to the full strength of the arguments offered by these scientists in support of their observations. It is interesting, though, to speculate on what would have happened had the measurements not agreed so well with the theoretical predictions. It is not clear whether such an event would have been regarded as an anomaly for the theory or as a doubtful measurement.

A somewhat different illustration of the complex interaction of observation and theoretical explanation is provided by the discovery of synchrotron radiation from Jupiter, or, to put it more accurately, the discovery that the microwave radiation emitted from Jupiter was due to synchrotron radiation. In 1959, Sloanaker[20] reported measurements of the intensity of radiation from Jupiter at a wavelength of about 10 cm that gave temperatures, computed from a black-body model, that ranged from 300°K to 1,010°K, with a mean of 640 ± 85°K. Those temperatures were inconsistent with temperatures obtained from earlier measurements at 3 cm and in the infrared region. An additional problem was that the considerable scatter in the data allowed a variable component in the radiation. At the same time, Drake and Hvatum[21] reported measurements at 22 cm and 68 cm that required temperatures of 3,000°K and 70,000°K, respectively, which they regarded as too high to be plausible. They combined their results with others and concluded that Jupiter was emitting nonthermal radiation. They proposed "that the radiation originates as synchrotron radiation from relativistic particles trapped in the Jovian magnetic field, a situation similar to the terrestrial Van Allen belts. A Jovian field of 5 gauss and a total number of particles 10^6 times greater than the terrestrial system will suffice to explain the observations."[22] There is at least a hint that the failure of the data to be consistent with the black-body model, combined with the variation and scatter in the data themselves, cast some doubt on the measurements.

Field then attempted to provide a consistent theoretical explanation of the observations.[23] He noted the failure of the black-body model and considered four other possible explanations: (1) thermal emission from deep in Jupiter's atmosphere, (2) emission in the ionosphere, (3) synchrotron radiation from relativistic electrons, and (4) cyclotron emission by nonrelativistic electrons, trapped in Jupiter's magnetic field. The first two were eliminated because they disagreed with the observations. The third was rejected because it required a density of high-energy electrons much greater than that seen near the earth, which Field regarded as implausible. "It seems therefore, that, on the basis of energy considerations, a belt of electrons trapped in a 1000 gauss field and replenished by the solar corpuscular emission could account for the decimeter radiation."[24] A second paper[25] refined the model and gave reasonably good agreement with the observations.

The situation was resolved after further measurements on decametric radiation, wavelengths of the order of 10 m, and a new calculation by Warwick,[26] based on the synchrotron radiation model. One problem for Field's model had been the rather large magnetic field required, approximately 1,000 gauss (G), which was regarded as unlikely. New measurements had suggested that the field had to be even larger, about 10,000 G. In addition, new polarization measurements seemed to favor the synchrotron radiation model, although the evidence was not conclusive. Warwick's model explained both the decimetric radiation and decametric radiation on the basis of interaction of external particles with the ionosphere of Jupiter. The electron densities required were approximately the same for both ranges of radiation, although a spectrum of energies was needed. The Jovian magnetic field required was only about 10 G. In addition, this model, which depended on external particles from the sun, explained the apparent correlation between Jupiter's radiation and the observed solar activity. It is interesting, then, that the currently accepted model agrees with Warwick's original proposal, except that it is believed that the electrons are locally generated and do not come directly from the sun.

This is more than simply a case in which a theoretical explanation provided additional confidence in a set of measurements. The observations helped to decide between the competing explanations, and the chosen explanation helped to validate what had been a confusing, and perhaps somewhat doubtful, set of observations. This sort of complex relationship between theory and observation happens frequently in physics.

This is not to say that observations are made in such a way as to agree with theory, but only that theoretical explanations rationally strengthen our belief both in our observations and in other observations made using the same technique for which we have no such theoretical explanation.

Yet another strategy is illustrated by observations of electric discharges in the rings of Saturn (SED) that were recorded during the flybys of Voyager 1[27] and Voyager 2.[28] If all plausible sources of error and all alternative explanations can be eliminated, then the observation is valid.

One possible explanation of the observations was poor data quality in the telemetry link between the spacecraft and earth. That was ruled out by the measured quality of the link (a form of calibration, to be discussed later) and by the fact that the observations were independent of the telemetry mode, a highly unlikely occurrence if the cause was a defect in the telemetry. Alternative explanations included discharges generated near the spacecraft through deleterious environmental phenomena. Such effects are seen when spacecraft are in the earth's atmosphere. No such discharges were seen when the spacecraft was near Jupiter, arguing that this explanation was not correct. The possibility of dustlike particles interacting with the spacecraft and causing the events was eliminated by the fact that the observed time scale of the discharges was far longer than that expected from dust interactions. The fact that the observations occurred when Voyager 1 was outside the magnetosphere of Saturn also argued against any interaction with the plasma or particulate environment of Saturn. The SED were also observed below a frequency of 100 kHz. That indicated that there was no ionosphere between Voyager 1 and the SED source, because the cutoff frequency for Saturn, below which signals are reflected, is 1.370 MHz. The observed peak in electron density (ionosphere) was far above Saturn's clouds, indicating that they could not be the source of the SED. After those studies were published, it was suggested that the rings of Saturn eclipsed part of the atmosphere and that for that region reflections would not occur,[29] and thus the argument was suspect. For Voyager 2, different parts of the atmosphere were eclipsed, and the discharges were again observed, arguing against the eclipse explanation. Having eliminated sources of error and alternative explanations, the conclusion was that "We believe, therefore, that the most probable source of SED is in Saturn's rings."[30]

The observation of Saturn electric discharges was confirmed by Voyager 2. Events with the same duration and frequency distribution were seen. Thus, the character of the events allowed the same arguments as given previously for Voyager 1. In addition, "Voyager 2 detected SED more than 48 hours before the first inbound bow shock crossing, so they clearly are not due to phenomena occurring at the spacecraft inside Saturn's magnetosphere."[31]

There were, however, some significant differences between the two sets of observations. Voyager 2 detected SED at a rate only one-third of that for Voyager 1. "In addition, the episodes themselves were distributed much more symmetrically about the time of closest approach (most of the Voyager 1 episodes occurred after the Voyager 1 encounter)."[32] Significant differences in the polarizations of the events were also observed. That led to the conclusion that "although Voyager 2 confirmed the SED phenomenon, the striking differences in polarization, episode distribution, and number of events strongly suggest a source that changes with time."[33] The character of the events allowed the use of the same arguments as to the existence of SED. The differences were then reasonably attributed to a changing source.

Perhaps the most widely used strategy for validation of results is that of calibration and experimental checks. Here the argument is that the ability of the apparatus to reproduce already known phenomena argues both for its proper operation and in favor of the results obtained. Calibration not only provides a check on the operation of the apparatus but also provides a numerical scale for measurement of the quantity involved. An example of this is the calibration of the Princeton experiment detecting $K_2^0 \rightarrow \pi^+ \pi^-$ and thus demonstrating the existence of CP violation, as discussed in Chapter 3.[34] In order to demonstrate the ability of their experimental apparatus to detect the decays $K_2^0 \rightarrow \pi^+ \pi^-$, they detected the already known phenomenon of regenerating K_1^0 mesons from K_2^0 mesons (K_1^0 also decay to $\pi^+ \pi^-$). Their procedure was described as follows:

An important calibration of the apparatus and data reduction system was afforded by observing the decays of K_1^0 mesons produced by coherent regeneration in 43 gm/cm^2 of tungsten. Because the K_1^0 mesons produced by coherent regeneration have the same momentum and direction as the K_2^0 beam, the K_1^0 decay simulates the direct decay of the K_2^0 into two pions. The regenerator was successively placed at intervals of 11 in along the region of the beam sensed by the detector to approximate the spatial distribution of the K_2^0's. The K_1^0 vector momenta peaked about the forward direction with a standard deviation of 3.4 ± 0.3 milliradians. The mass distribution of these events was fitted to a Gaussian with an average mass 498.1 ± 0.4 MeV and standard deviation of 3.6 ± 0.2 MeV.[35]

The agreement with the known K^0 mass of 497.7 MeV and with the expected angle of 0° argued for the ability of the apparatus to detect two-pion decays of the K^0.

They then compared their sample of suggested K_2^0 decays, those for which cos θ > 0.99999, with the regenerated K_1^0 decays:

The average of the distribution of masses of those events with cos θ > 0.99999 is found to be 499.1 ± 0.8 MeV. A corresponding calculation has been made for the tungsten data resulting in a mean mass of 498.1 ± 0.4. The difference is 1.0 ± 0.9 MeV. Alternately we may take the mass of the K^0 to be known and compute the mass of the secondaries for two-body decay. Again restricting our attention to those events with cosθ > 0.99999 and assuming one of the secondaries to be a pion, the mass of the other particle is determined to be 137.4 ± 1.8. Fitted to a Gaussian shape the forward peak in Fig. 3 has a standard deviation of 4.0 ± 0.7 milliradians to be compared with 3.4 ± 0.3 milliradians for the tungsten. *The events from the He gas appear identical with those from the coherent regeneration in tungsten in both mass and angular spread.*[36]

The agreement of the two sets of data gave confidence in their result.[37]

In any well-designed experiment there will be checks to ensure uniform, if not proper, operation of the apparatus.[38] An example of this is the careful monitoring of event rate as a function of K^0 intensity in the Princeton experiment discussed earlier. Failure to perform these checks, or performing them incorrectly, can result in invalid or erroneous results. An interesting case in point concerns one of the tests of time-reversal invariance. Time-reversal invariance requires that the angular distributions (at the same energy) in the reactions $\gamma + d \rightarrow n + p$ and $n + p \rightarrow \gamma + d$ be identical. The angular distribution for the first reaction had already been well measured, and the test involved looking at the angular distribution of the second reaction. The efficiency as a function of angle was needed so that the correct distribution could be calculated from the raw data. The experimenters used the γ rays resulting from $n + p \rightarrow d + \pi^0$, in which the π^0 decays into two γ rays, to measure that efficiency. That was done, and on the basis of that efficiency the two distributions differed significantly.[39] The conclusion that time-reversal invariance was violated, a dramatic result, was announced at a meeting of the American Physical Society.[40] Unfortunately, as a sophisticated Monte Carlo calculation showed later, the measured efficiency using the π^0 reaction was not correct. The calculation thus served as a check on the efficiency. The proper efficiency was used in the published version, and the final result showed no difference between the angular distributions, and time-reversal symmetry was still preserved.[41]

An experimental apparatus should also be able to detect artifacts that are known in advance to be present. This is a somewhat different kind of experimental check or calibration. An example of this comes from infrared spectroscopy of organic molecules.[42] It was not always possible to prepare a pure sample of such material. Sometimes one had to put the substance in an oil paste or in solution. In such cases, one expects to observe, superimposed on the spectrum of the substance, the spectrum of the oil or the solvent, which one can compare with the spectrum of oil or solvent alone. This is illustrated in Figures 6.1 and 6.2, where the absorption lines due to oil and $CHCl_3$, a solvent, are shown. Observation of this artifact gives confidence in the other observations, and if the spectrum of oil or solvent has been measured independently, it also provides a calibration of the apparatus.

An interesting sidelight occurred in the report of these observations. An unexpected background was observed (Figure 6.3; note absorption labeled as background):

Owing to what appears to have been a thin film of unknown nature deposited on the optical surfaces of the instrument as a result of an accident during its operation, there are three weak absorptions in these records which have no significance at any time since they were always present, even when no sample was placed in the beam.[43]

The artifactual nature of the absorptions was clear, because they appeared when no sample was present. One might, however, be tempted to skepticism[44] and use this background to cast doubt on the apparatus and on the other observations. That would not be correct:

The correctness of the explanation of the source of these bands was seen after the instrument was taken apart and cleaned, shortly following

178 *The neglect of experiment*

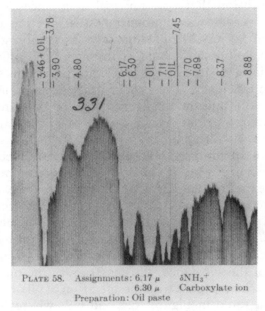

Figure 6.1. Infrared spectrum of an organic molecule prepared in an oil paste. The oil spectrum is clearly indicated. From Randall et al.[42]

the making of these records. After the cleaning, the spurious absorptions disappeared completely.[45]

This suggests another strategy. If an effect disappears when one predicts it will, then it is valid. This is, in fact, a special case of the strategy discussed earlier, in which observation of predicted behavior helps to validate a result, but it is so dramatic that I believe that it deserves its own discussion.

The "crucial" experiment[46] of Wu and associates[47] showing non-conservation of parity illustrates this very well. In that experiment, Co^{60} nuclei were polarized, and the decay electrons were detected. The asymmetry observed when the nuclei were polarized in two opposite directions demonstrated violation of parity conservation. The polarization could occur only at extremely low temperatures. One expected that as the temperature increased, the asymmetry would disappear. This is shown quite dramatically in Figure 6.4 and helps to validate the observation.[48] The warm counting rates, with no polarization, were independent of the

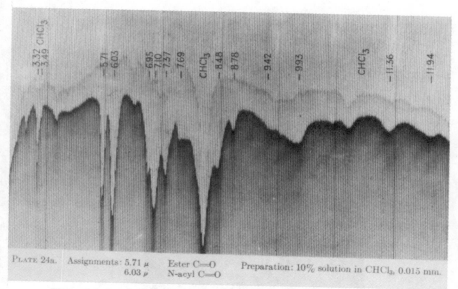

Figure 6.2. Infrared spectrum of an organic molecule in CHCl₃ solution. The CHCl₃ spectrum is shown. From Randall et al.[42]

magnetic-field direction, arguing against any instrumental asymmetry.

If, however, an effect does not disappear when it is predicted to, then one doubts both the proper operation of the apparatus and its results. This is illustrated in an experiment to measure the spin, or magnetic moment, of helium–6 nuclei using a Stern-Gerlach apparatus.[49] The He⁶ nuclei passed through a long, in-homogeneous magnetic field. If the spin was 1, then a decrease in the counting rate by a factor of three should have been observed when the magnetic field was on, as compared with when it was off. That was observed in an early data run. That would have been a startling result. Both accepted theory and all previous experimental work had shown that all even–even nuclei, of which He⁶ is an example, have spin zero, although this had never been directly tested for He⁶. In order to check on proper operation of the apparatus, the collimation of the beam was destroyed. Under those circumstances, no difference should have been observed between the field-off and field-on counting rates. The factor-of-three difference persisted and indicated that the magnetic field

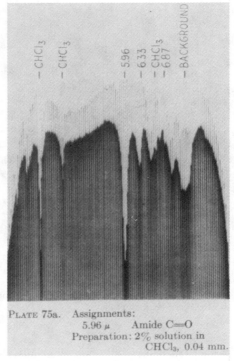

– CHCl3
– CHCl3
– 5.96
– 6.33
– CHCl3
– 6.87
– BACKGROUND
PLATE 75a. Assignments:
5.96 μ Amide C=O
Preparation: 2% solution in
CHCl3, 0.04 mm.

Figure 6.3 Infrared spectrum of an organic molecule in $CHCl_3$ solution. The $CHCl_3$ spectrum and the unexpected background absorption are shown. From Randall et al.[42]

was reducing the efficiency of the detector. It was only an unfortunate coincidence that the reduction was by the predicted factor of three. Subsequently the detector was more adequately shielded from the magnetic field. The final result showed no difference in counting rates, or spin zero. The importance of experimental checks is clear.

Another example of this is provided by the electron microscopic observations of the structures of myokinase and protamine by Ottensmeyer.[50] The electron micrographs he obtained are shown in Figures 6.5 and 6.6 (parts a and b) and do seem to indicate a structure. Dubochet[51] repeated those observations and after some difficulty obtained similar results for myokinase, shown in Figure 6.7 (parts 1–3). However, he also obtained similar micrographs

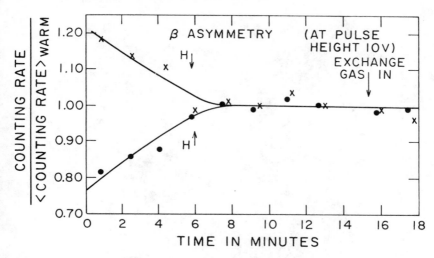

Figure 6.4. Decay electron counting rate as a function of time for decay of polarized Co^{60} nuclei. The asymmetry between the two polarizations is shown. From Wu et al.[47]

even when no protamine or myokinase was present, casting doubt on both sets of observations [Figures 6.7 (parts 4–6) and Figure 6.6 (parts 7 and 8)]. As he remarked, "These results demonstrate that a trained observer is astonishingly good in selecting structures in noise."[52] He suggested that other methods must be used to determine these structures.

Statistical arguments can also be an important part of validating a result, or in establishing that a particular effect was seen. Thus, the difference between the π^+ and π^- lifetimes divided by their average lifetime, a difference of 0.053 ± 0.068 percent, is consistent with zero difference, and the CPT theorem remains valid. Recall also the comments of Samuel Goudsmit, the editor of *Physical Review*, concerning Telegdi's results on parity violation (see Chapter 1). He argued that the original result, 0.062 ± 0.027, which was only a little more than two S.D., was not sufficient to establish the effect. He compared it to the originally reported asymmetry in η decay of 0.072 ± 0.028, in which later experiments showed that there was no asymmetry. He also contrasted it to "the overwhelming and compelling evidence" presented in the other two experiments, which, as we have seen, yielded 13 and 22 S.D.

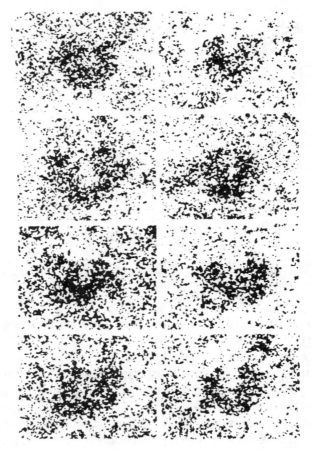

Figure 6.5. Electron micrograph of myokinase. From Ottensmeyer et al.[50]

An interesting point arose in the 1960s when the search for new particles and resonances using bubble-chamber techniques occupied a substantial fraction of the time of those working in experimental high-energy physics. The usual technique was to plot the number of events as a function of invariant mass or the square of invariant mass and look for bumps above a smooth background. The usual informal criterion for a new particle was that it give a 3-S.D. effect above background, which had a probability of 0.27 percent. That was extremely unlikely in any single experiment, but Professor Arthur Rosenfeld of the University of California is

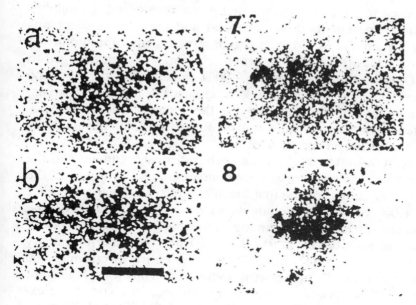

Figure 6.6. Electron micrographs (a, b) of salmon sperm protamine. From Ottensmeyer et al.[50] Electron micrographs (7, 8) obtained by Dubochet[51] with no protamine present. From Dubochet.[51]

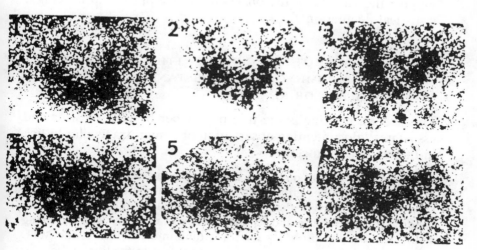

Figure 6.7. Electron micrographs (1–3) of myokinase obtained by Dubochet.[51] Electron micrographs (4–6) obtained by Dubochet with no myokinase present. From Dubochet.[51]

supposed to have pointed out that if one considered the number of such experiments done each year and the number of graphs drawn, then one would expect to observe a significant number of such 3-S.D. effects. The informal criterion was then changed to 4 S.D., whose probability was 0.0064 percent. The story may be apocryphal, but it did have wide circulation among physicists. Examination of the literature for the 1960s reveals several reported particles that were not confirmed by later observations and are no longer mentioned.

An interesting strategy using both statistical argument and the theoretical explanation of the phenomenon is illustrated in the following. Let us assume that the mass spectrum discussed earlier has 1,000 bins. The probability of observing 3 S.D. from the expected distribution, assuming the deviation in the number of events in each bin is statistical, is 0.27 percent. The probability of observing a single 3-S.D. effect in the entire spectrum (1,000 bins) is quite high; in fact, it is 93 percent. Thus, observation of such an effect would not be strong evidence in favor of the existence of a particle. If, however, a theory T predicted the existence of a particle at a mass m_0 and the 3-S.D. effect was seen in that particular bin, then not only would the observation provide evidence in support of the theory, but the supported theory would support the validity of the observation, as opposed to it being a statistical fluctuation.[53]

6.2 MICROTRABECULAR LATTICE OF THE CYTOPLASMIC GROUND SUBSTANCE: ARTIFACT OR REALITY?

This section heading is the title of a paper by Keith Porter and associates[54] documenting observation of a microtrabecular lattice in the cytoplasm in cells:

Published observations on the microtrabecular lattice (MTL)[55] that comprises the cytoplasmic ground substance (or matrix) have not met with universal acceptance; rather, they have stimulated questions as to the validity of the structure and its equivalence to what exists in the living cell. For now, and probably for some time into the future, the MTL will be regarded in some quarters as an artifact of fixation and/or dehydration, as merely a condensation of soluble proteins on elements of the so-called cytoskeleton. That it has been observed in frozen-dried cells and in others prepared by freeze-substitution as well as in cells after exposure to a variety of chemical fixations seems not to have been noticed. At the very least, these facts have not overcome the skepticism that new and somewhat unexpected findings of this nature usually stimulate.[56]

It is precisely because the observation was controversial and because the authors were aware of the importance and difficulty of establishing its validity that their epistemological strategies are of such interest. A summary of these strategies is contained in the abstract of the first paper:

The possibility that the lattice structure is an artifact of specimen preparation has been tested by (a) subjecting whole cultured cells (WI–38, NRK, chick embryo fibroblasts) to various chemical (aldehydes, osmium tetroxide) and nonchemical (freezing) fixation schedules, (b) examination of model systems (erythrocytes, protein solutions), (c) substantiating the reliability of critical-point drying, and (d) comparing images of whole cells with conventionally prepared (plastic-embedded) cells. The lattice structure is preserved by chemical and nonchemical fixation, though alterations in ultrastructure can occur especially after prolonged exposure to osmium tetroxide. The critical-point method for drying specimens appears to be reliable as is the freeze-drying method. The discrepancies between images of plastic-embedded and sectioned cells, and images of whole, critical-point dried cells appear to be related, in part, to the electron-scattering properties of the embedding resin. The described observations indicate that the microtrabecular lattice seen in electron micrographs closely represents the nonrandom structure of the cytoplasmic ground substance of living cultured cells.[57]

Method (a) established the result by appealing to "different" experiments within the same general technique (i.e., different cells, different methods of fixing). Here the differences seem to be of a more pragmatic nature (although certainly not theory-independent) than the theoretical differences discussed earlier. Although some differences were noted after prolonged exposure to osmium tetroxide, it had been established earlier, in an independent experiment performed by another experimenter, that that was to be expected. The importance of "different" methods was further emphasized in a note added in proof:

In the time that has elapsed since this paper was initially submitted for publication, several other cell types (platelets, myoblasts, thyroid epithelial cells, neutrophiles, and neuroblastoma cells) have been observed.

In each instance the MTL is present but with variations in form with cell type that are striking.[58] [K.R. Porter and K. Anderson, unpublished observations].[59]

We see again, as previously discussed for Saturn electric discharges, that some differences were attributed to different causes. The arguments in support of the existence of the MTL apply to all of the cell types, and it was regarded as established. The variations were then associated with the different types of cells and did not cast doubt on the lattice. The argument using different cells not only favored the existence of a lattice but also supported the view that it was a general feature of cells, not just of a particular type of cell. If only a single fixing method had been used, one might argue that this was a consistent artifact caused by the fixing. The fact that different fixing methods were used on different cells makes this implausible.

Examination of model systems (b) predicted that if the lattice was a precipitation artifact, it should have been observed in erythrocytes and in bovine serum albumin. The fact that no such structures were seen in those systems argued against the lattice being an artifact. The argument here was a prediction that the effect would not be seen under certain conditions, as discussed earlier. If the effect had appeared, that would have argued for it being an artifact. They further argued[60] that other cell structures revealed by electron microscopy, including those frequently questioned, had been shown to exist before fixation. They pointed out that those structures, which included ribosomes, microtubules in ordered arrangements, and microfilaments, were observed both by electron microscopy and by other methods, including light optical methods and cytochemical procedures. They concluded:

These several examples of fine details of morphology are readily accepted. Thus it is reasonable to believe that structures of similar dimensions, for several reasons not previously observed, are also faithfully preserved by glutaraldehyde–osmium tetroxide fixation.[61]

This is an example of indirect validation, as discussed earlier. Their final argument was as follows:

Perhaps most compelling in convincing us that the system under observation exists in the living cell is a study on its structural changes during cell function. The rapid transitions displayed by the lattice during the aggregation and dispersion of pigment in chromatophores are dramatic and cannot be interpreted as a product of fixation without an extraordinary strain on the imagination. The same must be said for the observed changes which take place in response to low temperatures. Although not so sensitive to chilling as the contained microtubules, the MTL deforms as though certain bonds within its structure are weakened while others are retained. This restructing of the lattice when normal temperatures are restored is both prompt and dramatic.[62]

They were arguing that the behavior of the lattice under certain circumstances was extremely implausible if it was an artifact, in similarity to the discussion of the moons of Jupiter given earlier. They were also "intervening" by varying the temperature and observing the predicted behavior of the lattice.

Despite the rather strong evidence presented for existence of the lattice, doubts remained (the quotation regarding the lack of universal acceptance is from a later paper). A second paper described a similar set of experiments in which most of the observations were derived from a single type of cell, PtK_2, from a male marsupial. This was done so that the influences of the preparation procedures could be compared more directly. They stated that "The morphologies after these various procedures are quite similar. All show the characteristic three-dimensional lattice or mesh work of slender filaments called microtrabeculae."[63] The similar structures are shown in Figure 6.8 for four different preparation procedures.

That paper also presented evidence for quantitative differences in the lengths and widths of trabeculae in cells that were first fixed with glutaraldehyde and then frozen and dried, as compared with cells that were simply frozen and dried (Figure 6.9). They attributed these differences to the already demonstrated (in an independent experiment) failure of glutaraldehyde to penetrate the cell quickly and fix any component in active motion. The question why the lattice had not been seen previously was also discussed:

Why has the MTL not been observed in electron micrographs of thin sections of the numerous cells reported on in the literature? It has been, but not with the same clarity that characterizes its appearance in thick sections from which embedding material has been removed or in whole cells. The reasons are that thin sections, because of their thinness, contain only fragments of the microtrabeculae, and parts of these are rendered invisible because they share with the epoxy matrix the same or similar electron scattering properties.[64]

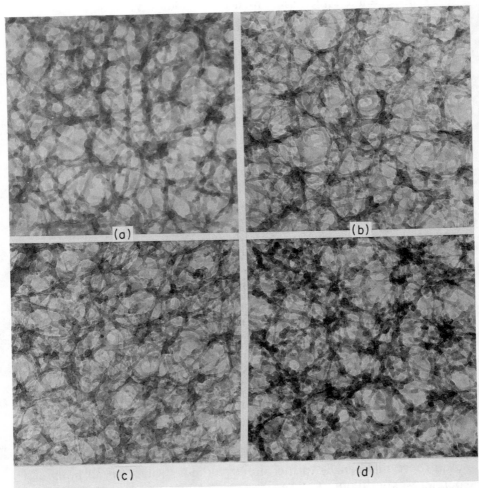

Figure 6.8. Electron micrographs of PtK$_2$ cells showing similar structures, presumably the MTL prepared by different techniques: (a) freeze dried; (b) fixed with glutaraldehyde, postfixed with OsO$_4$, and freeze dried; (c) freeze substitution; (d) fixed with glutaraldehyde, postfixed with OsO$_4$, and critical-point-dried. (Magnification is x108,000.) From Porter and Anderson.[54]

They concluded: "The essential message from these experiments is that the cytomatrix is structured."[65]

The issue still seems to be unresolved. Although most scientists in the field accept the idea that the cytoplasm has a complex, linked structure, there is still no universal agreement on the MTL

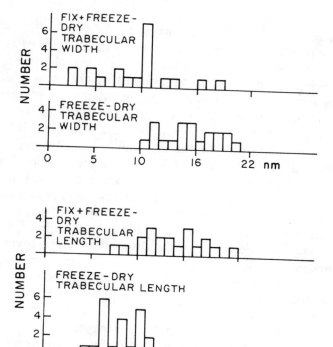

Figure 6.9. Lengths and widths of trabeculae prepared under different conditions. From Porter and Anderson[54].

as a unit structure.[66] As Professor Porter and his collaborators noted, "The trend away from the cytosol concept (a bag of enzymes) may by now have advanced to the point where the idea is largely an interest of history."[67] In this later paper, more detailed evidence for the lattice is presented from the behavior of chromatophores, discussed earlier. In particular, the experimenters "intervened" to both induce and stop the aggregation and dispersion of pigment. The attitude of the proponents of the MTL has changed from trying to present convincing or decisive evidence to the attitude that, "These few examples may suffice to relieve a few anxieties as to the equivalence of the MTL to the in vivo structure of the cytomatrix or hyalin cytoplasm."[68] The subsequent history should prove very interesting.

6.3   DISCUSSION

In this chapter I have presented a set of strategies that I believe provide grounds for rational belief in experimental results. Several of these strategies use the argument that observation of predicted behavior, or of known phenomena, argues in favor of proper operation of the apparatus, and this validates the observations or measurements. These include the following: intervention, illustrated by observations of changes in cell function and structure as the cell is cooled to low temperature; experimental checks and calibration, as shown in measurements of regenerated K_1^0 mesons as a check that the apparatus can measure K_2^0 decays, and observation of the spectrum of oil or solvent, as expected, in infrared spectroscopy of organic molecules; disappearance of an effect when predicted, as in the disappearance of asymmetry in the decay of polarized nuclei as the sample warms up. As discussed earlier, failure to observe predicted behavior casts doubt on the observations or measurements, as illustrated by failure of the observed effect in the Stern-Gerlach experiment on He^6 to disappear when the beam was destroyed.

Other strategies include the use of different experiments to validate an observation, as illustrated in observations on different cells using different fixing techniques to show the existence of the microtrabecular lattice. It would be quite implausible if the same structures appeared under these very different circumstances and were found to be artifacts of the apparatus. Similarly, it may be very implausible for the data to be artifacts, as illustrated by the argument concerning Jupiter's moons and by Millikan's results. If one can eliminate all of the plausible sources of error and all alternative explanations, then an observation such as that of the electric discharges in the rings of Saturn is valid. One also has reason to believe in observations made with an apparatus based on a well-corroborated theory, such as the radiotelescope or the electron microscope. Support for the theory also supports the observation. A similar feedthrough of support occurs when observations are explained or predicted by a well-corroborated theory, as shown in the discovery of the W bosons and the discovery of synchrotron radiation from Jupiter.

I do not suggest that this set of strategies is either exclusive or exhaustive, nor do I believe that any of them or any subset is either a necessary condition or a sufficient condition for such rational belief. I do not believe such a general method exists. The fact that all of these strategies are illustrated by examples from the work of practicing scientists argues against those[69] who believe that rational argument plays little, if any, role in validation of such results. In particular, the cases of the microtrabecular lattice and Saturn electric discharges demonstrate clearly that scientists do provide such arguments. I am not suggesting that because scientists behave this way, that behavior has methodological significance. I believe that independent and reasonable justification has been given for these strategies. I do suggest, however, that rational argument plays the major role in this issue.

7

The epistemology of experiment: case studies

In the previous chapter I argued that there is an epistemology of experiment, a set of strategies that are used to provide rational belief in an experimental result. These strategies distinguish between a valid observation or measurement and an artifact created by the apparatus. One such strategy, suggested by Hacking,[1] is intervention. Hacking's example involves use of the microscope, with the experimenter "intervening" (i.e., staining, injecting fluid) and in other ways manipulating the object under observation. This is a special case of a more general strategy in which we predict what will be observed if the apparatus is working properly or is working as we expect it to. When the observation is made, we increase our belief in the proper operation of the apparatus and in its results. Hacking also suggests that we believe in an observation if it can be made using "different" microscopes (i.e., optical, polarizing, phase-contrast, etc.) or if it is independently confirmed. If the apparatus is based on a well-corroborated theory, we also have good reason to trust its results. Even without such a theory we can validate observations. Sometimes the phenomena themselves provide such evidence. The observations of the moons of Jupiter by Galileo were extremely unlikely to have been artifacts of the telescope, as suggested by Cremonini, because they appeared to represent a consistent planetary system and even obeyed Kepler's Third Law (R^3/T^2 = constant).

If all plausible sources of error and all alternative explanations can be eliminated, then an observation is valid. Perhaps the most widely used strategy for validation of results is that of calibration and experimental checks. The ability of an apparatus to reproduce already known phenomena argues both for its proper operation and in favor of the results obtained. If, however, such checks fail or the predicted behavior does not occur, then one may legitimately doubt the observations. The fact that an observation can be explained using currently accepted theory can also help to provide validation.

I argued that these strategies have plausible justification and, in addition, provided examples showing their use in actual scientific practice. In this chapter I shall examine in detail four important experimental episodes from the history of twentieth-century physics and show that those experiments made use of these strategies. It is precisely because those experiments were of such importance that the strategies used are of interest. The episodes, all of which were considered earlier, were: (1) parity nonconservation, which provided an example of a "crucial" set of experiments that decided the issue, (2) CP violation, which was an example of a "convincing" experiment, (3) the nondiscovery of parity nonconservation, in which experiments showing parity-violating effects were conducted, but whose significance was not realized, and (4) Millikan's famous oil-drop experiments, which both established the quantization of charge and measured its fundamental unit.

In the discussion of these episodes, I shall, to a large extent, allow the experimenters to speak for themselves, so that the importance of the epistemological issues and strategies will be clear, not an interpretation by a philosopher of science. Some of the discussion was presented earlier, but it will be given here in full detail so that the entire set of strategies can be seen.

7.1   PARITY NONCONSERVATION

The principle of parity conservation, or space-reflection symmetry, had been an accepted and strongly believed principle in modern physics. In 1956, following failure to solve the "θ–τ puzzle" (in which particles of the same mass and lifetime decayed into states of opposite parity) by conventional means, Lee and Yang suggested that perhaps parity was violated in the weak interactions. They suggested several experiments to test their conjecture. I shall discuss two of these, β decay of oriented nuclei, and the sequential decay $\pi \rightarrow \mu \rightarrow$ e, because those were the first experiments done and because they provided the crucial evidence for the physics community. Lee and Yang described those experiments as follows: "A relatively simple possibility is to measure the angular distribution of the electrons coming from the β decays of oriented nuclei. If θ is the angle between the orientation of the parent nucleus and the momentum of the electron; an asymmetry of distribution between θ and $180° - \theta$ constitutes an unequivocal proof that parity is not conserved in β decay."[2]

In the decay processes

$$\pi \rightarrow \mu + \nu, \tag{5}$$

$$\mu \rightarrow e + \nu + \nu, \tag{6}$$

starting from a π meson at rest, one could study the distribution of the angle θ between the μ-meson momentum and the electron momentum, the latter being in the center of mass of the μ meson. If parity is conserved in neither (5) nor (6), the distribution will not in general be identical for θ and $(180° - \theta)$. To understand this, consider first the orientation of the muon spin. If (5) violates parity conservation, the muon would be in general polarized in its direction of motion. In the subsequent decay (6), the angular distribution problem with respect to θ is therefore closely similar to the angular distribution problem of β rays from oriented nuclei, which we have discussed before.[3]

The first experiment was performed by Wu and associates.[4] It consisted of a layer of polarized Co^{60} nuclei and a single electron counter that was located in a direction either parallel to or antiparallel to the orientation of the nuclei. The direction of the Co^{60} polarization could be changed, and any variation in counting rate in the fixed electron counter could be observed (Figure 7.1). Their results (Figure 7.2) clearly show an asymmetry in the counting rate. They concluded: "If an asymmetry in the distribution between θ and $180° - \theta$ (where θ is the angle between the orientation of the parent nuclei and the momentum of the electrons) is observed, it provides unequivocal proof that parity is not conserved in β decay. This asymmetry has been observed in the case of oriented Co^{60}."[5]

We now examine the arguments given to support the view that the observation was due to asymmetric β decay of Co^{60} nuclei, not due to an experimental artifact or another cause. It had been determined in another experiment[6] that Co^{60} nuclei could be polarized by the method used (the Rose–Gorter method) and that the degree of polarization could be determined by observing the anisotropy of the γ rays emitted. The Wu experiment included

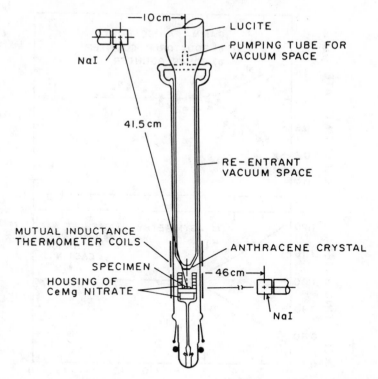

Figure 7.1. Schematic drawing of the experimental apparatus of Wu et al.[4] From Wu et al.[4]

two sodium iodide counters to detect the γ rays, one in the equatorial plane, and one near the polar direction, to measure anisotropy and thus monitor polarization (Figure 7.2). The ability of the apparatus to reproduce a known effect gave confidence in the results. In addition, the γ-ray counters were biased to avoid a known source of possible background, namely, γ rays from Compton scattering.

The electron counter, or β ray counter, was a crystal of anthracene, with a Lucite light pipe leading to a photomultiplier tube. The counter was calibrated by observing the known Cs^{137} conversion line. It was also checked for stability against magnetic and temperature effects. If the observed β asymmetry was due to polarized Co^{60} nuclei, the effect should disappear at the same time as the γ anisotropy. The experimenters remarked that "The

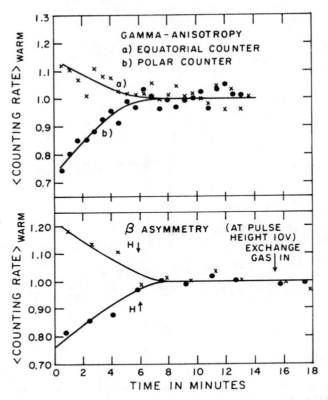

Figure 7.2. Gamma anisotropy and β asymmetry for polarizing field pointing up and pointing down. From Wu et al.[4]

time for disappearance of the beta asymmetry coincides well with that of the gamma anisotropy."[7] (See Figure 7.2.) If the two times had differed, then one might very well have questioned the observation. The fact that the warm counting rates (i.e., for no polarization) were independent of the polarizing-field directions argues against any instrumental asymmetry. It was also possible that the demagnetization field used to cool the sample might have left a remnant magnetization that caused the β asymmetry. This possible cause was eliminated by noting that the observed asymmetry did not change sign with reversal of the field direction. Another possible cause was the possibility of a small magnetic field perpendicular to the polarizing field due to the fact that the crystal axis was not parallel to the polarizing field.

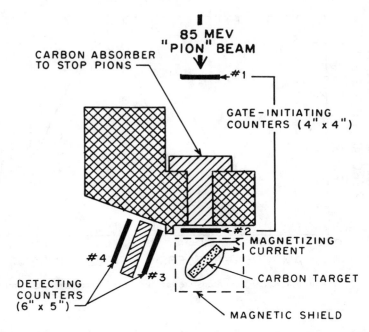

Figure 7.3. Experimental arrangement of Garwin et al.[9] From Garwin et al.[9]

To check whether the beta asymmetry could be caused by such a magnetic field distortion, we allowed a drop of $CoCl_2$ solution to dry on a thin plastic disk and cemented the disk to the bottom of the same housing. In this way the cobalt nuclei should not be cooled sufficiently to produce an appreciable nuclear polarization, whereas the housing will behave as before. The large beta asymmetry was not observed. Furthermore, to investigate possible internal magnetic effects on the paths of the electrons as they find their way to the surface of the crystal, we prepared another source by rubbing $CoCl_2$ solution on the surface of the cooling salt until a reasonable amount of the crystal was dissolved. We then allowed the solution to dry. No beta asymmetry was observed with this specimen.[8]

The second experiment, on $\pi \to \mu \to e$ decay, was performed by two different methods by Garwin, Lederman, and Weinrich[9] and by Friedman and Telegdi.[10] The experimental apparatus of Garwin and associates used a positive meson beam with an energy of 85 MeV from the Nevis cyclotron that contained approximately 10 percent muons (Figure 7.3). The beam passed through eight

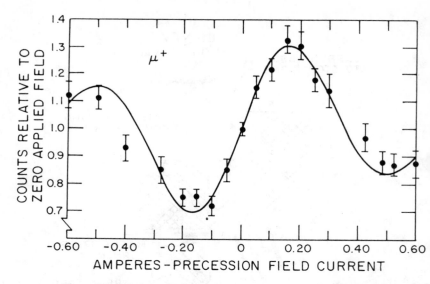

Figure 7.4. Relative counting rate as a function of precession field current. From Garwin et al.[9]

inches of carbon, eliminating the pions, whose range at 85 MeV is about five inches of carbon. The muons were stopped in a carbon target, and the decay electrons were detected in a telescope that restricted the electron energy to greater than 25 MeV and was at a fixed angle of 100° to the incoming beam. The stopped muon signal opened a gate 1.25 μsec long that was delayed 0.75 μsec and put in coincidence with the electron telescope. Thus, the telescope detected electrons with energy greater than 25 MeV, which were born between 0.75 μsec and 2 μsec after the muon stopped. The lifetime of the positive muon in carbon is approximately 2 μsec, and the electrons were due to muon decay.[11] The carbon target was placed inside a magnetic shield to guard against stray fields, and a small vertical magnetic field was applied. That field caused the muon spin to precess at a rate $(\mu/s\hbar)H$ radians per second. The decay distribution was carried along with the spin, so that the fixed counter could sample the entire distribution by looking at counting rate as a function of magnetizing current for a fixed time delay. A typical run is shown in Figure 7.4. The distribution would be constant, or flat, if parity was conserved. They concluded that parity was not conserved.

These authors noted that the apparatus had been used previously to measure the lifetimes of μ^+ and μ^- mesons "in a vast number of elements."[12] They also made a systematic check on the apparatus by reducing the absorber to five inches of carbon so that pions would stop in the target. They expected to see no asymmetry under these conditions, because the electrons would then come from muons emitted isotropically from pions decaying at rest in the carbon. They observed a factor of 10 increase in electron counting rate, as expected, and no variation in counting rate with magnetizing current. They also noted that "The only conceivable effect of magnetizing current is the precession of muon spins."[13] The asymmetry was also detected using different targets (nuclear emulsion, polyethylene, and carbon), and also with negative muons, giving additional confidence in the result.[14] "Proof of the 2π symmetry of the distribution and the sign of the moment was obtained by shifting the electron counters to 65° with respect to the incident muon direction. The repetition of the magnetizing run yielded a curve as in [Figure 7.4] but shifted to the right by 0.075 ampere (5.9 gauss) corresponding to a precession angle of 37°, in agreement with the spatial rotation of the counter system."[15] The fact that the curve shown in Figure 7.4 was a theoretical fit to a distribution of $1 - \frac{1}{3} \cos \theta$ calculated for the experimental apparatus used provided further confirmation. They also noted that a specific theoretical model proposed by Lee and Yang predicted, for their experimental arrangement, a peak-to-valley ratio of order 2.5 for electrons with energy greater than 35 MeV, compared with the measured value of 1.92 $\pm$ 0.19. A note added in proof stated that they had observed an energy dependence of a in the $1 + a \cos \theta$ distribution in rough qualitative agreement with that predicted by the two-component neutrino theory without derivative coupling.

The $\pi \rightarrow \mu \rightarrow$ e experiment was performed by a second method by Friedman and Telegdi.[16] A beam of π^+ mesons was stopped in nuclear emulsion. The decays $\pi^+ \rightarrow \mu^+ + \nu$ and the subsequent $\mu^+ \rightarrow e^+ + 2\nu$ can readily be observed. If parity is not conserved in the weak interactions, as discussed earlier, then the muons will be longitudinally polarized (i.e., along the direction of motion), and the decay electrons will show a θ and $\pi - \theta$ asymmetry, where θ is the angle between the muon initial direction and the electron momentum. The initial direction of the muon was chosen to avoid possible effects from multiple scattering in the emulsion. Multiple scattering might change the muon direction, but not its spin orientation.

Friedman and Telegdi published two papers on this subject:[17] a brief note giving their result, along with a short justification, and a longer paper giving more details of their analysis and experiment. As they said, "In view of the intrinsic importance of the subject, we consider it worthwhile to present our data at this preliminary stage."[18] In the early paper, the major problem discussed was the possibility of depolarizing effects masking or destroying a real parity-violating asymmetry. The first possible cause was the fringing field of the cyclotron, which would cause the muon spins to precess by different amounts, depending on their directions of emission. This is a different problem from that in the experiment of Garwin and associates, in which the magnetic field caused the aligned spins of all the stopped muons to precess by the same amount. Friedman and Telegdi used three concentric magnetic shields around the emulsion, which was then subject to a measured magnetic field less than or equal to 4×10^{-3} G. The precession due to such a field would be negligible. The second possibility was that the formation of muonium, an atomlike structure of μ^+e^-, would depolarize the muons. They stated that, "In the absence of specific experiments on muonium formation, one can perhaps be guided by analogous data on positronium $[e^+e^-]$ in solids."[19] This was a real effect, for, as mentioned earlier, and as stated in a footnote, Garwin and associates had found a smaller asymmetry in nuclear emulsion than in carbon. Such muonium formation could only mask a parity-violating effect, not cause it. They concluded that their observed asymmetry [(backward electrons − forward electrons)/Total] of 0.091 ± 0.022, based on 2,000 events,[20] was only a lower limit because of the possible depolarizing effects, but that their result still showed that parity was violated.

The second paper presented more details of the analysis. One possible source of bias was failure to observe all of the positrons from the muon decays. They found that in only 1 percent of the muon decays was the positron not detected, and that certainly could not account for the 9 percent asymmetry seen. They also stated that three different scanners supplied data and that each of their results was consistent with the mean: 0.06 ± 0.035, 0.11

± 0.036, and 0.13 ± 0.045. They performed an additional check on possible bias by determining the distribution of muon directions relative to the incident pion direction. They found that distribution to be isotropic, within statistics, as expected. The asymmetry observed was −0.026 ± 0.029.

7.2 CP VIOLATION

As discussed earlier, the experiments that demonstrated parity violation also showed violation of charge-conjugation invariance (particle-antiparticle symmetry). Even before the results of those experiments were known, Landau[21] had suggested that the symmetry of interest was "combined parity," or CP, the joint operation of charge conjugation and parity, and that CP would be found to be conserved in those experiments, even if C and P were separately violated. One consequence of that was that the K_1^0 and K_2^0 mesons became eigenstates of CP, with values +1 and −1, respectively. The K_1^0 decayed into two pions, but if CP was conserved, the K_2^0 meson could not decay into two pions. The experiment searching for $K_2^0 \rightarrow 2\pi$ was carried out by Christenson and associates.[22]

The experimental apparatus is shown in Figure 7.5. The products of a beam of K_2^0 mesons, decaying in a volume filled with helium, were detected in two spectrometers, each consisting of spark chambers, scintillation counters, magnet, and water Cerenkov counter. The scintillation counters guaranteed that a charged particle had passed through each spectrometer. The water Cerenkov counter ensured that the particle was an electron, muon, or pion. The spark chambers were triggered by coincidence between the counter telescopes on each side to make sure that the particles derived from a single K_2^0 decay. The spark chambers delineated the particle tracks in the magnetic fields, from which the vector momentum of each decay product could be calculated.

The invariant mass, m^*, was computed from the vector momenta, assuming each decay particle had the mass of a pion [$m^{*2} = (E_{\pi 1} + E_{\pi 2})^2 - (\mathbf{P}_{\pi 1} + \mathbf{P}_{\pi 2})^2$]. If both particles were indeed pions from K_2^0 decay, m^* would equal the K_2^0 mass. The vector sum of the two momenta allowed calculation of the angle θ between the sum and the direction of the K_2^0 beam. This angle should

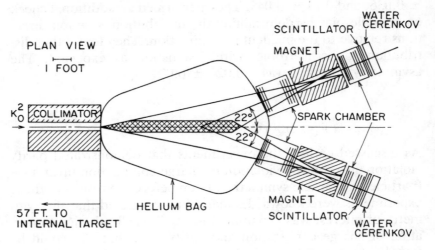

Figure 7.5 Plan view of the apparatus used by the Princeton group to detect $K_2^0 \rightarrow \pi^+ \pi^-$. From Christenson et al.[22]

be zero for two-body decays, but not, in general, for three-body decays.

A clear peak of 45 ± 9 events above the Monte Carlo calculation of expected background events for $\cos \theta \approx 1$ ($\theta \approx 0$) and m^* between 490 and 510 MeV/c^2 (the K_2^0 mass region) is shown in Figure 7.6b and Figure 7.7. If these are 2π decays of the K_2^0 meson, then CP symmetry is violated, and that was what the experimenters concluded.

They provided rather strong evidence in support of their conclusion. The Monte Carlo calculation was checked by comparing it to the observed m^* distribution for all events. The good agreement shown in Figure 7.6a indicates that the 2π events were not being simulated by other normal decay modes (i.e., $K_{\mu3}$, K_{e3}, $K_{\pi3}$). Events with $\cos \theta > 0.9995$ were remeasured on a more precise measuring machine and recomputed using an independent computer program. Those events are shown in Figure 7.7 for three mass regions: one above, one below, and one encompassing the mass of the neutral K meson. The peak near $\cos \theta = 1$ is seen only for the K^0 mass region. If it were an artifact, one would expect to see similar peaks in the other two regions.

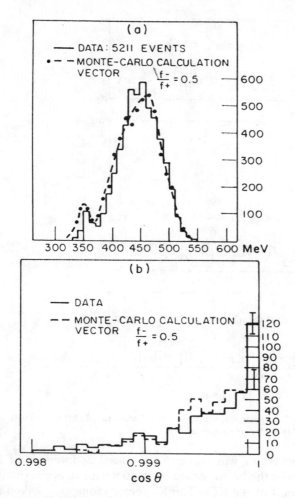

Figure 7.6.(a) Experimental distribution of m^* compared with the Monte Carlo calculation. The calculated distribution is normalized to the total number of events. (b) Angular distribution of events in the range $490 < m^* < 510$ MeV. The calculated curve is normalized to the number of events in the complete sample. From Christenson et al.[22]

An important calibration of the apparatus and data reduction system was afforded by observing the decays of K_1^0 mesons produced by coherent regeneration in 43 gm/cm^2 of tungsten. Since the K_1^0 mesons produced by coherent regeneration have the same momentum and direction as the K_2^0 beam, the K_1^0 decay simulates the direct decay of the K_2^0 into two

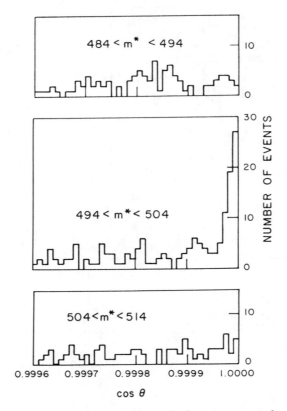

Figure 7.7. Angular distributions in three mass ranges for events with cos θ > 0.9995. From Christenson et al.[22]

pions. The regenerator was successively placed at intervals of 11 in. along the region of the beam sensed by the detector to approximate the spatial distribution of the K_2^0's. The K_1^0 vector momenta peaked about the forward direction with a standard deviation of 3.4 ± 0.3 milliradians. The mass distribution of these events was fitted to a Gaussian with an average mass 498.1 ± 0.4 MeV and standard deviation of 3.6 ± 0.2 MeV.[23]

The results from the regenerated K_1^0 events were compared with those from the presumed K_2^0 decay. For events with cos θ > 0.99999, the mass was found to be 499.1 ± 0.8 MeV/c^2. That was in good agreement with the value of 498.1 ± 0.4 MeV/c^2 from the regenerated events, the difference being 1.0 ± 0.9 MeV/c^2. A Gaussian fit to the forward peak gave 4.0 ± 0.7 milliradians, compared with 3.4 ± 0.3 milliradians for the tungsten events, again in good agreement. They also computed the mass of one secondary, assuming that the other was a pion, for events with $\cos \theta > 0.99999$. The mass was found to be 137.4 ± 1.8 MeV/c^2, in good agreement with the known pion mass of 139.6 MeV/c^2. They concluded: "*The events from the He gas appear identical with those from the coherent regeneration in tungsten in both mass and angular spread*" (emphasis in original).[24]

These experimenters also took data with a hydrogen target to look for the possibility of anomalously large regeneration, which had been reported previously. They found no such anomalous regeneration, and the number of events left after subtracting regeneration was consistent with the decay data. They also examined other possible explanations for the forward peak in the angular distribution at the K^0 mass and rejected them. Those included K_1^0 coherent regeneration in the helium, $K_{\mu 3}$ or K_{e3} decay, and decay into $\pi^+ \pi^- \gamma$. They stated that "any alternative explanation of the effect requires highly nonphysical behavior of the three body decay modes of K_2^0."[25]

Not all of the checks and calibrations were reported in the published paper. (This continues to be a problem today, because results of importance tend to be published in letters journals, which restrict the length of an article.) A detailed examination of the laboratory notebook for that experiment[26] shows that such checks were in fact carried out (e.g., calibrating a neutron monitor to measure K^0 intensity, checking that the K^0 picture-taking rate and other quantities of interest stayed constant as a function of K^0 intensity).

It is interesting to compare the epistemological strategy of the Princeton experiment[27] with that of Abashian and associates[28] published three weeks later. In the latter, events similar to those seen in the Princeton experiment were observed, albeit with lower statistics, and no claim was made that $K_2^0 \rightarrow 2\pi$ had been observed, much less that CP was violated.

The experiment of Abashian and associates was not, in fact, designed to look for $K_2^0 \rightarrow 2\pi$, but rather to look at $K_{\mu 3}^0$ ($K_2^0 \rightarrow \mu^\pm \pi^\mp \nu$) and K_{e3}^0 ($K_2^0 \rightarrow e^\pm \pi^\mp \nu$) decays. Two runs were taken, one with the trigger of the apparatus set for muons, and the other with an electron trigger. 4,500 V events, indicating neutral K^0 decay, were found, for a total of 9,000 events. Of the

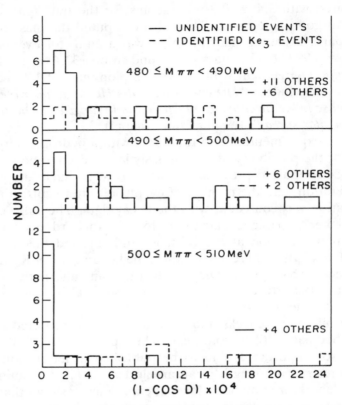

Figure 7.8. Angular distribution of reconstructed K_2^0 relative to beam direction for events having masses near 498 MeV when analyzed assuming the decay is $K_2^0 \rightarrow \pi^+ \pi^-$. K_{e3} run. From Abashian et al.[28]

muon trigger events, 1,700 were identifiable as true $K_{\mu3}$ events, whereas for the electron trigger, 750 K_{e3} events were identified. All of the 9,000 events were reanalyzed assuming that they were $K_2^0 \rightarrow \pi^+ \pi^-$. They also calculated the invariant mass and the angle D that the vector sum of the momenta made with the K^0 beam direction. "If the decay were truly a two-pion decay of the K_2^0 the mass would be the K_2^0 mass of 498 MeV and the angle D would be zero within the experimental angular resolution of about $0.5°$."[29] Their results for electron events are shown in Figure 7.8. A pronounced forward peak is seen for the unidentifiable events (no clear electron signal) in the bin 500 to 510 MeV, whereas a less pronounced peak was seen in the other bins. For identifiable

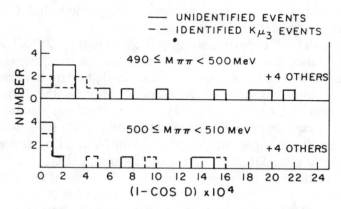

Figure 7.9. Angular distribution of reconstructed K_2^0 relative to beam direction for events having masses near 498 MeV when analyzed assuming the decay is $K_2^0 \rightarrow \pi^+\pi^-$. $K_{\mu 3}$ run. From Abashian et al.[28]

K_{e3} events, only a smooth distribution was observed. They also noted that the calibration of their magnetic field may have been in error, causing their reconstructed K^0 mass to be too high. Figure 7.9 shows similar distributions for muon events, including identified $K_{\mu 3}$ events. They stated that "Because of the meager statistics, no definite conclusions can be drawn from this data [muon run] except to say that no marked difference occurs between the unidentifiable and identifiable $K_{\mu 3}$ events. *We cannot rule out the possibility that the effect observed in the K_{e3} run is due to $K_{\mu 3}$ decay*" (emphasis added).[30]

They then examined the consequences of labeling a K_{e3} or a $K_{\mu 3}$ event as a $K \rightarrow 2\pi$ event. The K_{e3} events could not give a forward peak for masses near the K^0 mass. The $K_{\mu 3}$ decay could, however, give a small angle for D for masses near the K^0 mass. Thus the importance of the prior reservation that the events in the K_{e3} sample might be due to $K_{\mu 3}$ decay. They stated "Because of this behavior of the $\pi\mu\nu$ events selected in this way, and because of the possible peaking observed in the events of [Figure 7.9], we prefer to be very cautious about asserting that the pronounced peaking in the $500-510$ MeV bin of [Figure 7.8] indicates a $\pi\pi$ decay of the K_2^0 meson."[31] They concluded: "The peak in [Figure 7.8] constitutes about $0.2-0.3\%$ of all K_2^0 decays. We feel safe in assigning this as an upper limit. In terms of the $K_2^0 \rightarrow \pi^+\pi^-$ to the $K_1 \rightarrow \pi\pi$ rate this places an upper limit on CP violation of 5 x 10^{-6}, with at least a suggestion that CP may actually be violated by this amount."[32]

Although both experiments used the fact that "real" K_2^0 decays should reconstruct with the K^0 mass and have zero angle with the beam line, the Princeton experiment was able to demonstrate that for coherently regenerated K_1^0 decays, whereas the Abashian experiment did not have that available. That was because the Princeton experiment was designed to look for $K_2^0 \rightarrow 2\pi$, whereas it was an afterthought for the other group. The Princeton group was able to eliminate alternative decay modes, (K_{e3}, $K_{\mu3}$, etc.), but Abashian and associates could not, because of their higher statistics, 45 ± 9 as compared with 11 events for Abashian and associates. The importance of statistical considerations is clear. In addition, the Princeton group showed that anomalous regeneration could not explain the events, using their hydrogen data. The Abashian group could not demonstrate that.

7.3   THE NONDISCOVERY OF PARITY NONCONSERVATION

In the late 1920s and early 1930s there were experiments that showed parity-violating effects, but whose significance was not realized at the time. Those experiments involved double scattering of electrons from a β-decay source. As shown in Figure 7.10, longitudinally polarized electrons from β decay will give rise to asymmetry between the scattering at 90° and the scattering at 270°. I shall not deal here with the reasons why the significance of these experiments was not recognized, or why the observations were not repeated, which has been dealt with in Chapter 2. What makes this episode so interesting are the changing views concerning the observation of asymmetry. That observation was first thought to be valid, then discounted as an instrumental effect, then thought again to be correct, and later again incorrectly attributed to instrumental effects. That observation is now thought to be valid.

The first experiment was that of Cox, McIlwraith, and Kurrelmeyer.[33] The context of the experiment included the wave nature of electrons, recently discovered by Davisson and Germer, and the introduction of electron spin, by Goudsmit and Uhlenbeck. It was speculated that electron spin would act in a manner

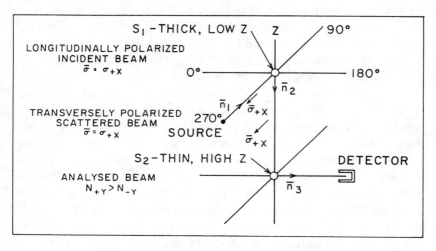

Figure 7.10. Double scattering for a longitudinally polarized incident beam. From Grodzins.[47]

Table 7.1. *Experimental results of Cox and associates* [33]

Count at 90° / Count at 270°	0.76	0.90	0.94	0.87	0.98	1.03	1.03	0.91
Probable error	0.01	0.07	0.01	0.02	0.01	0.03	0.02	0.02
Count at 90° / Count at 270°	0.095	0.99	1.01	1.06	1.05	0.55	0.91	
Probable error	0.05	0.03	0.04	0.05	0.02	0.05	0.03	

Note: weighted average 0.91 ± 0.01. (This average was not calculated in the original paper.)

similar to the transverse-electric-field vector in optical experiments and would result in the first scattering polarizing the electrons, which would be analyzed by the second scattering. The experiment is shown schematically in Figure 7.10, and the results are given in Table 7.1. The weighted average (counts at 90°)/(counts at 270°) = 0.91 ± 0.01. The experimenters stated that "Since they [the experiments] are soon to be interrupted it seems advisable to make a preliminary report of the results obtained thus far, although they are somewhat inconclusive in spite of a great accumulation of data."[34] The major sources of background in the experiment were photoelectrons ejected from the apparatus by γ rays also emitted by the radium. "The high penetration of these rays makes it impossible to shield against them without interposing so much material that the path of the β-particles would be too much lengthened. Their effect is considerably reduced by making the counting chamber of ebonite, from which comparatively few photo-electrons are ejected. Their number, however, could not be neglected, but there is no reason to expect that it would vary between the two settings at which most of the counts were made."[35] The platinum-point Geiger tubes were also unreliable. Not only did they give inconsistent results after an hour or two of operation, forcing their replacement, but different points gave different results. The experimenters took data so as to minimize these effects: "... the count at one setting (i.e. 90°) is compared with the counts immediately preceding and following it at the other setting. Thus, for example, from eleven counts five values are obtained for the ratio of the count at one setting to the count at the other. The mean of these values is taken and the probable error computed in the usual way. It is these means and probable errors that are given in the following table."[36] The results were then given, and the remainder of the paper was devoted to discussing the validity of the results. That portion will be presented in its entirety, despite its length, because the kinds of epistemological issues we have been discussing were addressed directly, although in somewhat different language:

It will be noted that of these results a large part indicate a marked asymmetry in the sense already mentioned. The rest show no asymmetry beyond the order of the probable error. The wide divergence among the results calls for some explanation, and a suggestion to this end will be offered later. Meanwhile a few remarks may be made on the qualitative evidence of asymmetry. Since the apparatus is symmetrical in design as between the two settings at 90° and 270°, the source of the asymmetry must be looked for in an accidental asymmetry in construction or in some asymmetry in the electron itself. The following possibilities may be suggested in the former case. The radium and the point in the counter were doubtless not exactly centered. But they were removed and replaced repeatedly in the course of the observations, and it seems unlikely that their accidental dislocations could be so preponderantly in one direction as are the observations. Any effect due to this cause could be offset by turning the counter and the rod that carries the radium through 180°. The apparatus with which these data were obtained was not designed to make this convenient, but in the latest apparatus these rotations can be made without disturbing anything else. The results thus far obtained with this apparatus do not lead us to believe that this factor is effective. There was doubtless some asymmetry in the targets as regards their orientation and surface condition. These also were several times removed and replaced after their surfaces had been freshly filed bright. A magnetic field inside the cavity due either to the slight penetration of the earth's field through the steel walls or to an accidental magnetization of the cylinder itself would introduce an asymmetrical factor. It seems highly unlikely, however, that any deflection so caused could be great enough to produce effects of the magnitude observed. It seems possible, on the other hand, that a spinning electron might be oriented by even a weak field by a kind of space quantization and that this orientation might combine with the scattering to produce the observed asymmetry. This explanation, of course, assumes a polarity in the electron as definite as that required to explain the observations as due to double scattering. Of the same sort is the supposition that the beam of β-particles undergoes a polarization in passing through the mica windows, similar to the polarization of light in passing through a tourmaline crystal. This effect was in fact looked for carefully in an experiment auxiliary to the present investigation but it was not found.

It should be remarked of several of these suggested explanations of the observations that their acceptance would offer greater difficulties in accounting for the discrepancies among the different results than would the acceptance of the hypothesis that we have here a true polarization due to the double scattering of asymmetrical electrons. This later hypothesis seems the most tenable at the present time. The discrepancies observed we ascribe tentatively to a selective action in the platinum points, whereby some points register only the slower β-particles. Observations in apparent agreement with this assumption have recently been made by N. Riehl. It is necessary to suppose further that the polarization is also selective, the effect being manifest only in the faster β-particles. In support of this it may be remarked that a few observations we have just made seem to show that asymmetry is more consistently observed when a piece of celluloid or cellophane is placed in front of the counter to stop the slower β-particles. Perhaps the simplest assumption here is that only β-particles which are scattered without loss of energy show polarization.[37]

That work was continued by Carl Chase, then a graduate student working for Richard Cox. He noted the previous work of Cox and associates and stated that "The results were not consistent, but most of the runs showed an effect of this nature."[38]

Table 7.2. *Chase's 1929 results* [38]

Azimuthal angle	Relative count
0°	1.000
90°	0.977
180°	0.958
270°	0.969

The first apparatus he used was very similar to that of Cox, except that the upper target could be turned out of the beam to allow a count of the γ rays alone. "A few runs showed, in the total count, an asymmetry of the same nature as had been observed by the earlier three, the effect being in the same direction and on the average of the same magnitude."[39] Chase then took runs in which the γ-ray counts and the total counts (electrons plus γ rays) were measured. These were of the same order of magnitude, with the γ-ray counts sometimes being larger. The apparatus was then modified to get more electron counts per γ-ray count. The milligram radium source was replaced by one that contained primarily radium E, and which gave off fewer γ rays. A hard rubber lining was placed in the lower part of the apparatus to absorb the photoelectrons, and the apparatus had thinner windows for the electrons. That proved quite successful: "The total count was thus reduced to scattered electrons; when the upper target was turned through 180° so that no electrons can be scattered to the lower target, all counts disappear, and reappear immediately the target is returned to its usual position."[40] This also allowed the counters to be run at lower voltages, with more consistency and longer lifetimes (approximately twelve hours or more).

Chase's 1929 results are shown in Table 7.2. "It will be seen that there is no indication of any effect of polarization of the kind suspected by Cox, McIlwraith and Kurrelmeyer.... One must conclude that a beam of electrons is not polarized by scattering."[41] He also found no explanation for the asymmetry seen by Cox and seen in his early runs, except possibly that of γ rays, which would have required an apparatus asymmetry that was not apparent. The fact that he counted only electrons gave him confidence in his result.

Chase continued his work and began investigating the properties of platinum-point Geiger tubes.[42] He surveyed the literature and found that among the double-scattering experiments, those with slow electrons had given negative results, whereas some of those with fast electrons had shown polarization effects. In that 1930 paper he reported on the velocity dependence of Geiger-tube efficiency. He found, using a magnetic spectrometer, that at low voltages more slow electrons were counted, whereas at higher voltages more fast electrons were counted. That result was checked by placing a paper absorber in the electron beam, a maneuver expected to reduce the number of low-velocity electrons, as determined by magnetic deflection, but to have no effect on higher-velocity electrons. That effect was observed. Chase further checked that the counts at high voltage were due to electrons and were not spurious: "It may be noted that besides computing the velocities by measurement of the magnetic field strength, the spectrum was measured with a gold leaf electroscope, and found to be in agreement with the photographic work of Curie and d'Espine."[43] The validity of the result was established by the agreement among three different measurement methods.

Chase concluded that his new work cast doubt on his previous experiment which had failed to confirm the results of Cox and associates: "From the results contained in this paper it appears that this later work [his previous paper] did not include a count of the faster electrons. Thinner windows allowed more slow electrons to get through. The increased number of electrons in turn allowed the operation of the counter at lower voltages, making for more consistent results, but counting only the slower electrons. . . . The foregoing shows the need for further work on these experiments, which we have undertaken."[44] Preliminary experiments indicated that a gold-leaf electroscope made a good electron detector, and that was what Chase adopted for further work.

Chase reported that work two weeks later. The apparatus was similar to that used previously, except that a gold-leaf electroscope was used as a detector. "The electroscope is far more sensitive than the point discharge counters which we have used, to say nothing of its infinitely greater reliability and ease of use. The one disadvantage, that it is very sensitive to gamma-rays, can be eliminated by proper shielding."[45] Fast electrons were obtained by using radium E as a source, with slower electrons eliminated

Table 7.3. *Experimental results of Chase*[45] *(15 September 1930), relative counts*

At 0°	At 90°	At180°	At 270°	Weight
1.000	0.972	1.009	1.024	1
1.000	0.975	1.075	1.075	1
1.000	0.997	0.986	1.005	1
1.000	0.990	0.986	1.015	1
1.000	0.988	1.000	1.008	1
1.000	0.994	0.976	1.010	1
1.000	1.034	1.041	1.044	1
1.000		0.950		4
	1.000		1.030	3
	1.000		1.040	3
		1.000	1.020	2
1.000		0.933		1
	1.000		1.030	2
		1.000	0.969	2
1.000			1.003	1
1.000		1.037		2
	1.000	0.933		2
				Weighted
1.000 ± 0.003	0.993 ± 0.003	0.985 ± 0.003	1.021 ± 0.003	means

Note: experimental error = 1%.

by absorbing screens in front of the electroscope. His results are given in Table 7.3 and clearly show the 90°–270° asymmetry: "The asymmetry between the counts at 90° and 270° is always observed, which was in no sense true before.... As an interesting sort of check, the apparatus which had previously given a negative result was set up again; with counters used as they had been before, at lower voltages; the results were negative as before, but with high voltages on the counter, high enough to ruin the point within an hour or two, the effect was very likely to appear. Making no changes except in the voltage of the counter, the effect could be accentuated or suppressed."[46] Chase did express some reservations about the experiment at the very end of the paper. These reservations involved backgrounds due to secondary γ rays and secondary electrons and the inability of the apparatus to distinguish between primary and secondary scattered electrons, but he did not cast any doubt on his experimental results.

In all fairness, as I reported in Chapter 2, there is a problem with these experimental results. Results from modern experiments and theoretical calculations agree that there are more electrons observed at 90° than at 270°, contrary to the results of Cox and Chase. Can those earlier results be spurious, or are they due to confusion concerning the coordinate system?[47] I believe the latter is correct. In addition to the arguments given in each paper by the experimenters, there is the fact that the effect was seen with two rather different experimental apparatuses. Perhaps the most telling point is that in both experiments, the correct velocity dependence was observed. The polarization of the β-decay electrons increases with increasing velocity, and that is what was seen. It would indeed be remarkable, and very improbable, if a spurious effect could give the same velocity dependence in two different experiments. The subsequent history of those experiments[48] showed many repetitions of the scattering at 0° and 180°, which tested Mott's calculation and Dirac's theory. Chase seems initially to have regarded Cox's results as valid, but he changed his mind after his results on slow electrons, and then attributed Cox's results to an instrumental effect. He then reversed his opinion, on the basis of his later work with the electroscope as a detector. The results of both Chase and Cox were then treated as valid, but were neglected because of the seemingly theoretically more important experiments on 0° − 180° scattering. They were later regarded as being due to instrumental or apparatus effects, and were rediscovered and regarded as correct only following the discovery of parity violation in 1957 and repetitions of the experiments.

7.4 MILLIKAN'S OIL-DROP EXPERIMENTS

Millikan's oil-drop experiments are justly famous because they established the quantization of charge and also measured the fundamental unit of charge to a precision an order of magnitude better than any previous method. It is because Millikan's method of measuring charge was novel and also claimed so much greater precision than any previous measurements that his arguments both for the validity of his observations and for their accuracy and precision are of interest. In the following discussion I shall deal with both of these questions.

Gerald Holton has pointed out[49] that in Millikan's series of measurements of e, his method changed. He began by using H. A. Wilson's modification of J. J. Thomson's cloud method, in which a cloud produced in an ionized fog chamber by sudden expansion was allowed to fall freely under gravity, and its velocity was measured. A second, presumably similar, cloud was subjected to an electric field superposed on gravity, and its velocity was measured. From those two velocities, the charge e could be determined. In the course of improving this method, Millikan discovered a new method involving observations on a single water drop that avoided many of the difficulties of the cloud method.[50]

One of those difficulties was the problem of determining the temperature in the chamber, so that the correct viscosity of air, which is a function of temperature, could be used in calculation of e. In the cloud method, the temperature was assumed to be the equilibrium temperature after condensation, "a quantity obtained from theoretical considerations relating to the adiabatic expansion of saturated vapours and the experimental curve expressing the relation between the temperature and density of saturated vapour."[51] Millikan measured the mean temperature of the chamber using a standard mercury manometer and found that the calculated temperature was 14.2°C, whereas the measured temperature was 26°C. He further refined his measurement by using a thermocouple to measure the air temperature at the point of observation. By an independent experiment, he found that the couple had a small enough thermal capacity to respond with very little lag to changes in the temperature of the surrounding gas.

This conclusion was further confirmed by comparing the observed and computed values of the instantaneous fall in temperature produced by an adiabatic expansion of the air in the carboy. The computed fall was 1°.16C., the observed fall was 1°.12C. The couple was next placed inside the fog chamber, . . . and an expansion of 20 cm. of mercury produced. The galvanometer showed a sudden movement of 2mm., and within ten seconds had crept back to its original zero. One degree of change in temperature was found by a separate experiment to produce a deflexion of 8.5mm. Hence the maximum fall in temperature in the expansion as indicated by this couple was about ¼°C.[52]

That result, along with tests using another thermocouple, convinced Millikan that the temperature in the chamber was approximately room temperature. He then used that temperature to determine a value for the viscosity of air in the chamber, which differed from Wilson's value by about 10 percent. The problem of determining the correct value for the viscosity of air would remain with Millikan throughout those experiments.

Millikan had planned to try to balance the cloud with an electric field, so that its entire evaporation history would be observed and the error due to evaporation eliminated: "It was not found possible to balance the cloud as had been originally planned, but it was found possible to do something very much better; namely to hold individual charged drops suspended by the field for periods varying from 30 to 60 seconds."[53] The problem of evaporation still remained, although Millikan believed that the observations would allow him to adequately correct for it.

Millikan also had an experimental check on the problem of evaporation. He measured the fall of a drop as it crossed three crosshairs. If there was significant evaporation, the time required to fall through the bottom space would be greater than that for the top space. That was not observed in general, indicating that evaporation was not a significant problem. That new method also eliminated the problem of having to perform measurements on two successive clouds, with the assumption that they were identical. Henceforth, all measurements could be performed on a single water drop. "There is no theoretical uncertainty whatever left in the method unless it be an uncertainty as to whether or not Stokes' law applies to the rate of fall of these drops under gravity."[54] Millikan pointed out that the validity of the law for large spheres had been established in an experiment done by H. S. Allen and that for his spherical drops, which were thirty-five to fifty times the mean free path of the air molecules, "it is scarcely conceivable that Stokes' law fails to hold for them."[55] He also noted that if there were a correction to Stokes's Law, his results would not agree with those of Rutherford.

Millikan further justified his observations by noting that they had been taken by two different observers, M. Begeman and himself, and that neither had had any idea of what the correct time of fall should be. He also regarded the fact that virtually all of his data had been used as supporting his result. Although he varied the weight assigned to each series of observations according to his subjective judgment, he excluded only six uncertain observations, which, he noted, also gave results in good agreement with his other observations. "In the third place I have discarded one uncertain and unduplicated observation apparently upon a singly charged drop which gave a value of the charge on the drop some 30 per cent lower than the final value of *e*."[56] That he attributed to rapid evaporation. He also varied his experiment by using alcohol drops in one series of observations and obtained consistent results.

Millikan also believed that his results themselves argued for their validity:

It will be observed that the only possible elementary charge of which the observed charges are multiples is 4.65 x 10^{-10} and further that the measured charges represent all the possible multiples of this charge between 2 and 6. This shows that the elementary charge cannot possibly be the smallest charge we observed: namely 9.3 x 10^{-10}, since we obtained odd as well as even multiples of half this quantity. Furthermore the elementary charge can scarcely be a submultiple (e.g. a half) of 4.65 x 10^{-10}, since the observed charges would then represent only even multiples of this elementary charge, and there is no reason why odd as well as even multiples of the elementary charge should not have been observed.[57]

Millikan further validated his results by showing that they were consistent with four other determinations of *e*, by different methods, those of Planck, Rutherford and Geiger, Regener, and Begeman, although he did not include the work of Ehrenhaft, de Broglie, Perrin, and Moreau in his final mean, because of experimental uncertainties, which he discussed. Of particular importance for the accuracy and precision of his result was Begeman's similar value obtained using Wilson's cloud method. He noted that the chief source of error in his earlier measurement, with Begeman, using the cloud method had been in the wrong value for the viscosity of air. The concordance of the two new measurements gave confidence in the new value for that viscosity. Although Millikan quoted no error for his value of *e*, he did state that the experimental uncertainties were reduced to those in the time measurement. He estimated the error of a single measurement as no more than two parts in fifty and noted that the error in the mean of a considerable number of concordant observations would be much less.

Millikan's next major improvement in the experiment[58] was to substitute oil drops for water drops and thereby solve the problem of evaporation. In addition to evaporation, he listed three possible sources of error and uncertainty in his previous work: (1) motion of the air, (2) lack of electric-field uniformity, and (3) the assumption of exact validity of Stokes's Law. The first two problems were dealt with by observing the oil drops within a closed chamber and by having the plates and spacers machined to high precision to ensure electric-field uniformity.

It was found, however, that values of e determined from drops of different sizes were different. That he attributed to a failure of Stokes's Law, which was corrected by substituting $F = F_{stokes}/$ $(1 + Al/a)$, where l is the mean free path of air molecules, a is the radius of the oil drop, and A is a constant. By plotting $e_1^{2/3}$ (where e_1 is the calculated charge on the drop) against l/a, the constant A could be determined empirically. He found that his value agreed with a theoretical calculation done by Cunningham based on kinetic theory considerations, although Millikan emphasized the empirical nature of the correction. Once that correction was made, consistent values of e were obtained for drops of all sizes.

Millikan supported his view that evaporation and air disturbances had been eliminated by presenting data on a drop that had been observed for a period of 4.5 hours, as compared with the 30- to 60-sec observations made on water drops: "How completely the errors arising from evaporation, convection currents or any sort of disturbances in the air are eliminated is shown by the constancy during all this time in the value of the velocity under gravity. . . . Further evidence of the complete stagnancy of the air is furnished by the fact that for an hour or more at a time the drop would not drift more than two or three millimeters to one side or the other of the point at which it entered the field."[59] As Millikan stated, the velocity under gravity did change slightly over the 4.5 hours. The time taken to fall 1.010 cm changed from 22.28 sec to 23.46 sec. That posed no great difficulty for Millikan, because all other quantities of interest were measured at the same time as the velocity under gravity (i.e., velocity with field on, voltage), so that e could be determined for each value of the time taken to fall under gravity.

To validate his measurement, Millikan presented arguments similar to the ones he had used previously. He noted that the observations had been taken by himself and Mr. Harvey Fletcher, on hundreds of drops with total charges varying from 1 to 150, and the drops had been composed of different substances, such as oil, mercury, and glycerine. He "found in every case the original charge on the drop an exact multiple of the smallest charge which we found that the drop caught from the air. The total number of changes which we have observed would be between one and two thousand, and *in not one instance has there been any change which did not represent the advent upon the drop of one definite invariable quantity of electricity or a very small multiple of that quantity*" (emphasis in original).[60] The consistency of the results forms part of the basis for Millikan's belief in his result. Millikan was using the change in the charge on the drop as the primary evidence for charge quantization. The value of e was determined from the total charge on the drop. He presented his final data, involving thirty-three drops, which he claimed were all the drops, save three, experimented on during forty-seven consecutive days.[61] Those three, which gave values of e 2 to 4 percent low, were attributed to two drops stuck together because of dust. Millikan noted that before he had made provisions to eliminate dust there had been many drops of that kind. In determining his final value of e, Millikan also excluded the four slowest and four fastest drops: "These are omitted not because their introduction would change the final value of e, which as a matter of fact isn't appreciably affected thereby, but solely because of the experimental uncertainties involved in work upon either exceedingly slow or exceedingly fast drops. When the velocities are very small residual convection currents and Brownian movements introduce errors, and when they are very large the time determination becomes unreliable, so that it is scarcely legitimate to include such observations in the final mean."[62]

Millikan's final result for e was 4.901×10^{-10} esu. A footnote[63] corrected the value to 4.891×10^{-10} esu, because of a better calibration of the voltmeter, which reduced the value by 0.06 percent, and an error in his calculation, in which he used the real mass instead of the apparent mass, which includes the buoyant effects of air, which resulted in a further reduction of 0.14 percent. Millikan cited no error in his final value, although he did note that the statistical error in the mean was 0.04 percent, and other measurement errors were of the order of a few tenths of a percent. He compared his value of e with that in the recent work of Regener, who reported a value of 4.79×10^{-10} esu with a probable

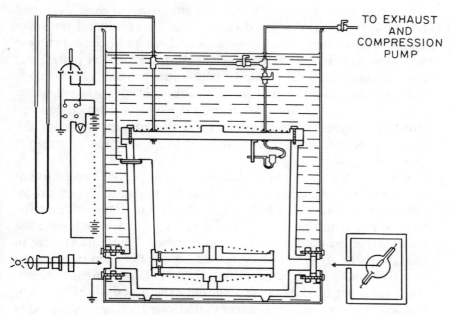

Figure 7.11. Millikan's oil drop apparatus. From Millikan.[64]

error of 3 percent, and noted that the two results agreed within the limits of observational error. He also stated that the only results that were in disagreement with his were those of Ehrenhaft and Przibram, and he offered a detailed criticism of their method.

More serious for Millikan was the fact that his new value of e, 4.891×10^{-10} esu, differed by almost 5 percent (Millikan stated 4 percent) from his previously reported value of 4.65×10^{-10}. When the new Stokes's Law correction was applied to the older value, he reported that the difference increased to 8 percent. He had stated that the previous timing uncertainty was no more than 2 or 3 percent. He attributed the rest of the difference to convection currents and other disturbances in the air that would appear in an expansion chamber, which was used in the earlier measurements, and to a lack of uniformity in the electric field in such an experiment. Both of these effects, he stated, would tend to reduce the value of e.

Millikan's final oil drop experiment was published in 1913.[64] The apparatus is shown in Figure 7.11. It should be emphasized that in that paper Millikan was concerned primarily with precise measurement of e. He regarded the quantization issue as settled by his 1911 work, discussed earlier. He made several improvements on his earlier apparatus: Dust, which had been a problem with approximately 10 percent of the drops in his 1911 paper, was almost completely eliminated. The voltmeter and chronoscope (time measurement) were carefully calibrated so as to have errors of no more than 0.1 and 0.2 percent, respectively. He also devoted substantial effort to determination of the correct value for the viscosity of air. Convection currents were also more carefully guarded against.

It was in the 1913 paper that Millikan reported two different methods for calculating e from his data. His earlier method involved using the sum of the time of fall under gravity and the time of rise with the electric field on ($1/t_g + 1/t_f$), which gave the total charge on the drop. A second method, using the change in charge on the drop, involved the difference between two successive times with the field on ($1/t_f' - 1/t_f$). The agreement between these different methods of calculating e increased Millikan's confidence in his result. That agreement also argued against the idea that distortion of the drop by the field modified the law of motion of the oil drop through the medium. "Now a very careful experimental study of the relations of $(1/t_g + 1/t_f)_0$ and $(1/t_f' - 1/t_f')_0$ shows so perfect agreement that no effect of distortion in changing measurably the value of e can be admitted."[65] Millikan claimed, however, that in his determination of e, only the first method of calculation was used.[66]

In the 1913 paper, Millikan also reported concordant observations using drops of different materials (oil, glycerine, and mercury), although the final value of e was determined only from observations on oil drops. The consistency of all his observations added to Millikan's confidence in his result. "*It will be seen that there is but one drop in the 58 whose departure from the line amounts to as much as 0.5 per cent. It is to be remarked, too, that this is not a selected group of drops but represents all of the drops experimented upon during 60 consecutive days* [emphasis in original], during which time the apparatus was taken down several times and set up anew."[67]

Millikan's final value for e, $(4.774 \pm 0.009) \times 10^{-10}$ esu,[68] was determined from only twenty-three of his fifty-eight published drops, those for which the correction to Stokes's Law was less than 6 percent, so that even if his empirical correction to the law was slightly in error, the effect on *e* would have been negligible.

The statistical error in the mean was 16 parts in 61,000, and the final error included uncertainties in the viscosity of air, the distance between the condenser plates, the voltmeter readings, and the distance between the crosshairs. Millikan estimated each of those four to be, at most, 0.1 percent. Thus his final error of 0.2 percent. Millikan noted that the error in his time measurements, previously a major source of error, had become negligible compared with the other uncertainties. Millikan compared his final results for *e* and *N*, Avogadro's number, to those determined by Planck's radiation method, Regener's radioactive method, and the Brownian motion method. He estimated the errors in these values as 3 percent and calculated that "The mean results by each one of the three other methods fall well within this limit of the value found above by the oil drop method."[69]

Millikan's new value of $e = (4.774 \pm 0.009) \times 10^{-10}$ esu was an embarrassment when compared with his 1911 value of 4.891×10^{-10} esu. Millikan attributed the difference to changes in the value of η, the viscosity of air, and in *A*, the empirical correction to Stokes's Law, both of which tended to lower the value of *e*, although by only a very small amount. The discrepancy was attributed primarily to the fact that the earlier distance-of-fall measurement was in error because the electric field had not been parallel to the gravitational field.

This distance could be measured with great accuracy but *the procedure assumed that the drop remained exactly at this distance throughout the whole of any observation, sometimes of several hours duration* [emphasis in original]. But if there was the slightest lack of parallelism between gravity and the lines of the electric field the drop would be obliged to drift slowly, and always in the same direction, away from this position, and a drift of 5 mm. was enough to introduce an error of 1 per cent. Such a drift could in no way be noticed by the observer if it took place in the line of sight, for the speeds of the drops were changing very slowly anyway because of evaporation, fall in the potential of the battery, etc. and a change in time due to such a drift would be completely masked by other causes of change. This source of uncertainty was well recognized at the time of the earlier observations and steps were taken at the beginning of the present work to eliminate it. It was in fact responsible for an error of nearly two percent.[70]

Millikan had modified his optical system for viewing the drops so that the depth of focus "was so small that a motion of ½ mm. blurred badly the image of the drop."[71] The alert reader will recall that Millikan had indeed worried about the same problem in his 1911 paper. In 1913 he was suggesting that the drop had drifted by almost 10 mm, in order to account for the discrepancy, whereas he had earlier remarked "that for an hour or more at a time the drop would not drift more than two or three millimeters to one side or the other of the point at which it entered the field."[72] It seems clear that one of Millikan's arguments for the accuracy of his result, either that of 1911 or that of 1913, was wrong. The point here is not to decide which value is correct (in fact, the difference between Millikan's 1913 value $e = 4.774 \times 10^{-10}$ esu and the modern value $e = 4.803242 \times 10^{-10}$ esu is due to a change in the value for the viscosity of air), but rather to examine the arguments that Millikan gave.

7.5   DISCUSSION

It seems clear that the strategies suggested in the previous chapter were indeed used in these four important experimental episodes. Millikan used different observers, different materials, different methods of calculation, and comparisons with the results of other experimenters using different methods to help validate his results. Calibration and checking of experimental apparatuses are illustrated in the use of regenerated K_1^0 mesons by Fitch and Cronin, and in the observations of Wu and associates, who used the previously determined γ-ray asymmetry to show polarization of nuclei. Garwin and associates intervened and moved their electron counter 35° and observed, as predicted, that the distribution shifted by that amount. Wu and associates also observed that their asymmetry disappeared as the sample warmed up, as expected. Millikan substituted oil drops for water drops to avoid the problem of evaporation, and Cox's method of taking data was explicitly designed to avoid a known problem with his Geiger counters. A well-corroborated theory of his apparatus certainly helped to give Millikan confidence in his results, particularly his agreement with Cunningham's theoretical calculation of the correction to Stokes's Law, although Millikan did test his own theoretical assumptions experimentally. The phenomena themselves helped to validate the results of Chase and Millikan and helped to cast some doubt on Cox's result. The expected effects were seen in all of Chase's runs and were claimed to have been seen with all of Millikan's drops, whereas Cox noted that his observed asymmetry was seen in only about half of his runs. The foregoing are but a few clear examples of these strategies being used in those studies. One strategy that has not been cited is that of using the theoretical explanation for an observation to help support the observation. All of the foregoing were episodes in which new phenomena were observed, calling for new theories. In the case of CP violation, no theoretical explanation is available even now. In the case of the observations of Cox and Chase, one may speculate that had they agreed with Mott's calculation, based on Dirac theory, they would have been more readily accepted.

The fact that scientists use these strategies does seem to argue in their favor, because, after all, science does appear to progress, and thus its methods must have some validity. I also argued in the last chapter that these strategies can be independently justified. Examination of four episodes, no matter how exemplary, does not, of course, demonstrate that these strategies are universally or even widely used. My own examination of the literature indicates that these are the strategies used by experimentalists, and the illustrative examples presented are all taken from the actual practice of physics. The contention that these strategies are in general use is supported by the cases presented, which deal with different experimenters, different types of apparatus, and different aspects of physics, over a period of more than fifty years.

8

Forging, cooking, trimming, and riding on the bandwagon: fraud in science

Another possible problem concerning the role of experiment in physics is that of scientific fraud. Accusations of fraud have recently received considerable publicity in both the scientific literature and popular literature.[1] Such accusations have raised serious questions concerning science policy and the funding of research, and several United States congressional committees have held hearings on the subject.[2]

More important for the philosophy of science, and science itself, are the doubts such accusations raise concerning the reliability of experimental data. If a large majority of scientists do not subscribe to Merton's norm of "disinterestedness,"[3] the honest reporting of experimental results, then science may be faced with an insoluble problem. Which experimental claims are to be believed – those that report a result or those that contradict it? Merton stated that "The virtual absence of fraud in the annals of science, ... appears exceptional when compared with the record of other spheres of activity...," but the accusations mentioned earlier may cast doubt on that assertion. Although no significant data on the prevalence of scientific fraud currently exist, Golley has expressed the view that these rather sensational cases are but "the tip of an iceberg."[4]

The question at issue here, however, is whether or not the normal procedures of science provide sufficient safeguards against fraud. It seems clear that some safeguards were operating in the cases noted earlier, because the fraud (or alleged fraud) was disclosed. In this chapter I shall examine four cases from the history of physics, involving not only possible fraud but also questionable analysis of data, to see if the existing procedures did work, and, if so, why. Two of these episodes, Rupp's double–scattering experiments in the 1930s (Chapter 2) and Millikan's oil-drop experiments (Chapter 5), have already been discussed in detail and will be briefly summarized here.

Before examining the case studies, it should be instructive to consider a taxonomy of different types of scientific fraud. In 1830, Charles Babbage, in his *Reflections on the Decline of Science in England*[5] (indicating that concern with scientific fraud is not a purely contemporary phenomenon), discussed three kinds of fraud – forging, cooking, and trimming – a list that seems to cover most cases. Forging is the recording of observations never made. Cooking includes selection of data to achieve agreement, and although Babbage did not discuss it, we should include in this category the tactic of selecting only those data that will agree with a particular hypothesis or theory. "Trimming consists in trimming off little bits here and there from those observations which differ most in excess from the mean."

8.1 RUPP'S EXPERIMENTS ON ELECTRON SCATTERING

Our first example of fraudulent data derives from Rupp's experiments on double scattering of electrons during the early 1930s. The context of Rupp's work was Mott's theoretical calculation,[6] based on Dirac's theory of the electron, which predicted that in double scattering of electrons, the first scattering would polarize the electrons, and the second scattering would analyze this polarization, resulting in asymmetry between the forward scattering and backward scattering. Mott specified the conditions under which this would be observed, including single, large angle scattering at high velocities from high-Z, or heavy, elements. Further work by Mott modified the calculations slightly and gave numerical results.[7]

Most experiments during that period showed no asymmetry, in disagreement with Mott's calculation, and cast doubt on Dirac's electron theory, already corroborated in atomic spectroscopy. A few, however, gave positive results. By far the most positive results were those reported by Rupp. It is fair to say that Rupp's observations showed the existence of the effect predicted by Mott, although the numerical agreement was not exact. There are, however, two important reasons for doubting the validity of Rupp's results. The first of these is Rupp's withdrawal, in 1935, of five papers, including two on electron scattering. A footnote cited a doctor's report that from 1932 onward Rupp had unknowingly suffered from a mental illness that had allowed aspects of a dream world to intrude into his work: "Examination showed that Dr. Rupp has suffered since 1932 from a state of mental weakness linked with psychogenic states of semi-consciousness (psychasthenia). During this illness and influenced by it, he published accounts of physical phenomena (positions, atomic disintegration), which had the character of 'fictions,' without being conscious of this. It is a matter of a breakthrough of dreamlike states into the field of his researches. There is no doubt that he will fully recover." Rupp also confessed his fraud. According to Louis de Broglie, "But Rupp confessed he had made up his experiments and a little later he went mad. Therefore the situation is not clear, and his experiments could not be repeated, which proves they are falsified."[8] In addition, later experimental and theoretical work done during the 1940s showed that under the conditions Rupp had used, his results could not have been obtained, in agreement with the other negative experimental results of that period. I have not been able to determine whether Rupp's fraud involved forging, cooking, or trimming. The only evidence I have been able to locate on this issue is purely anecdotal and indicates that Rupp's work may, in fact, have involved forging. H. R. Post (private communication) was told that after Rupp's withdrawal, his locked laboratory was opened, and no apparatus for electron scattering was found. What was found, according to another story, was apparatus for producing forged data (H. Frauenfelder, private communication). Professor W. Paul, a student at Berlin at the time, also heard this story (private communication). There is little doubt, however, given both Rupp's self-repudiation and later work by others, that his work was fraudulent. There is no plausible way that his results could have been honestly obtained. Whether this fraud was due to Rupp's mental illness or was deliberately perpetrated is not known. The question still at issue, however, is what effect Rupp's fraud had on the work of the physics community both before and after the fraud was revealed.

   In this case, as the detailed history presented in Chapter 2 shows, it seems clear that the normal procedures of the scientific community acted as a safeguard against Rupp's fraud. The history shows that an experiment that had important theoretical implications was repeated many times. Even prior to Rupp's self-repudiation, the community accepted the preponderance of experimental evidence in favor of the null result. This is, however, more than simply a case of the preponderance of evidence or of a wrong result being superseded by later and better work. Rupp's work was fraudulent. Widespread fraud would make science impossible, for one would then have no confidence in any experimental results. Thus, the defrauder can no longer be considered a member of the scientific community, and his work is no longer a part of the scientific literature.[9]

This episode also provides additional evidence, if it is needed, against a naive falsificationist account of science. The experimental results on double scattering of electrons were regarded as anomalous for Dirac's already well-corroborated electron theory. That theory was not rejected, but was modified in various ad hoc, although unsuccessful, ways. In addition, the experiment was repeated many times. One may speculate here that if the results had agreed with Dirac's theory, then so many repetitions would not have occurred. Work continued on the problem until the discrepancy was finally resolved.

8.2 MILLIKAN'S OIL-DROP EXPERIMENTS

In Millikan's famous 1913 paper, "On the Elementary Electrical Charge and the Avogadro Constant," he stated that the 58 drops under discussion had provided his entire set of data. "*It is to be remarked, too, that this is not a selected group of drops but represents all of the drops experimented upon during 60 consecutive days.*"[10] That statement is false. Millikan's laboratory notebook for that period shows that Millikan made observations from 28 October 1911 until 16 April 1912 and that he recorded data on 175 drops. Even if one were to count only observations made after 13 February 1912, the date of the first observation Millikan published, there would still be 49 excluded drops. Millikan also excluded observations within the data on a single drop and used selective calculational procedures.

As we discussed in detail in Chapter 5, Millikan's early exclusion of data was justified on the grounds that he was not sure his experimental apparatus was working properly. His exclusions of observations within a single drop had no major effect on his results, and Millikan did not, in fact, know the effect of that exclusion. During the period when Millikan was presumably confident of his apparatus, 49 drops were excluded. Millikan made no calculation of e for 22 of the 49 excluded drops. The most plausible explanation for their exclusion is that Millikan did not need them for his determination of e. As noted earlier, Millikan had far more data than he needed to improve the measurement of e by an order of magnitude. The 27 excluded events for which he did calculate e are more worrisome. Twelve were excluded because they required a second-order correction to Stokes's Law, two because of equipment malfunctions, five because they had few reliable observations, and two for no apparent reason, presumably because they were not needed.

Six drops remain to be explained. One was quite anomalous and will be discussed separately later. For each of the other five, Millikan not only calculated a value of e but also compared it with an expected value ("1% low"). His only evident reason for rejecting these five events is that their values did not agree with his expectations. The effect of excluding the five events under discussion was to make Millikan's data appear more consistent, to make the "largest departure from the mean value . . . 0.5%." Had he included those events, the departure would have been 2 percent. His mean value of $e = 4.778 \times 10^{-10}$ esu, for twenty-three events, would have changed to 4.784×10^{-10} esu, with a larger statistical error (see Table 8.1). This is a clear case of trimming. Babbage remarked that "This fraud is not perhaps so injurious . . . as cooking. The reason of this is, that the *average* given by the observations of the trimmer is the same. . . . His object is to gain a reputation for extreme accuracy in making observations,"[11] a comment that would seem to apply to Millikan.

Millikan's trimming also extended to his analysis of individual drops. Using Millikan's observations, one can calculate e by two methods: the first using the total charge on the drop, the second using the changes in charge on the drop. As Millikan noted,[12] the former method is more reliable, and he stated that he had used it exclusively. For 19 of 58 published drops, however, he used some combination of the two methods, usually the average, or the second method alone. The effects of this trimming were quite small, as shown in Table 8.1, but it did tend to make the results more consistent.

Table 8.1. *Comparison of Millikan's values (RAM) and Franklin's values (ADF) for* e

Drops	e (RAM) ($\times 10^{10}$ esu)	e (ADF) ($\times 10^{10}$ esu)	σ (RAM)[a]	σ (ADF)
Published				
First 23[b]	4.778	4.773	±0.002	±0.004
All 58	4.780	4.777	±0.002	±0.003
First 23[c]	4.776	4.773	±0.003	±0.004
All 58[c]	4.778	4.777	±0.002	±0.003
First 23[d]	4.784	4.782	±0.005	±0.006
All 58[d]	4.782	4.781	±0.003	±0.003
Unpublished				
To 2/13/1912		4.750		±0.010
After 2/13/1912		4.789		±0.007
Almost all drops[e]	4.781	4.780	±0.003	±0.003

[a]Statistical error in the mean.
[b]Events used by Millikan to determine *e*.
[c]With problematic drops corrected as in Table 5.2 (p. 154).
[d]With problematic drops corrected and with the excluded drops.
[e]The published 58, corrected, plus 25 unpublished measured after 13 February 1912.

More serious is Millikan's exclusion of the anomalous drop mentioned earlier. That event, the second drop of 16 April 1912, was among Millikan's very best observations. It had a large number of measurements, and the two methods of calculating *e* gave results that were consistent both internally and with each other. Millikan liked it: "Publish. Fine for showing two methods. . . ." When Millikan calculated *e* for that event, he found a value some 40 percent lower than his other values. He dismissed the event with the comment "Won't work" and did not publish it. Millikan may have excluded that event to avoid giving Ehrenhaft ammunition in the charge-quantization controversy. This would seem to be a case of cooking. In Millikan's defense, we may note that although there are no obvious experimental difficulties with the event, the data require that not only the total charge on the drop but also each change in that charge be an integral multiple of a fractional charge, a highly unlikely event.

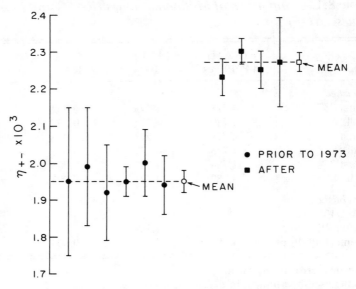

Figure 8.1. Measurements of $|\eta_{+-}|$ in the order of their publication.

The effect of Millikan's trimming was quite small, as we have seen. The effect of his cooking, which was, of course, unknown to the physics community, was also small. It certainly did not discourage Ehrenhaft, who continued his work through the 1920s. Nor would the exclusion of that event have had any significant effect on the searches for fractional charge since the 1960s. We may also note that this experiment has been repeated numerous times, with no observation of fractional charge.

8.3   THE PECULIAR HISTORY[13] OF $|\eta_{+-}|$

In 1964, the violation of CP (charge-conjugation parity) invariance was discovered by Christenson, Cronin, Fitch, and Turlay.[14] The parameter that measures the degree of violation $|\eta_{+-}|$ was measured six times prior to 1973, with a mean of $(1.95 \pm 0.03) \times 10^{-3}$ (Figure 8.1). The confidence level in the mean had increased with each succeeding measurement and stood higher than 99 percent after the six measurements. In 1973 and 1974, however, two further measurements were reported[15] that gave values of $(2.30 \pm 0.035) \times 10^{-3}$ and $(2.23 \pm 0.05) \times 10^{-3}$, in clear disagreement with the earlier experiments. Since then, two further measurements have been made that agree with the 1973 and 1974 measurements. The mean for the four latest measurements is $(2.27 \pm 0.022) \times 10^{-3}$ (Figure 8.1).[16] The means for the two sets of measurements differ by 8 S.D., an extremely improbable result if they were all measurements of the same quantity. Unless one is willing to consider seriously the idea that the value of $|\eta_{+-}|$ changed in 1972,[17] a suggestion few scientists or philosophers of science would entertain, then there is a serious, and seemingly systematic, discrepancy between the two sets of measurements.

There is certainly no suggestion of fraud here, but the episode does raise the possibility of a bandwagon effect, in which later measurements of a quantity tend to agree with earlier ones, and that will be discussed later. In addition, both of these sets of experiments (and those to be discussed in the next section) were high-energy physics experiments carried out by large groups of experimenters. Fraud would have required a conspiracy by at least several of the experimenters. We may also note that during such experiments, which involve long running times at accelerators, the group is usually asked to present progress reports and preliminary results at seminars, in part to justify continued access to the beam. Such scrutiny makes fraud even more unlikely.

It is interesting to look at the reasons why the post-1973 measurements were made, because the evidence indicated that $|\eta_{+-}|$ was already well measured. The reasons for doing the first post-1973 experiment were stated clearly by Professor Jack Steinberger, one of the group leaders: "When we first proposed this experiment, we took it for granted that a more precise measurement of ϕ_{+-} [the phase of the CP-violating amplitude] might have given a clue on the origin of CP violation, still one of the outstanding problems. This was the physics motivation for constructing the detector. There was another purely experimental: we saw a way of doing a much better measurement than had been done."[18] At the time, ϕ_{+-} had not been measured accurately. However, in order to measure ϕ_{+-}, one must use an interference technique that involves both the magnitude and the phase of the amplitude, so that in a sense the measurement of $|\eta_{+-}|$ is free. There is also some prospect of future utility in more precise determination of a quantity, because a theory change may, at some point, make the value of that quantity important.

The second post-1973 remeasurement of $|\eta_{+-}|$ was done only after the experimenters knew of the preliminary results of Steinberger experiment.[19] The original purpose of this second post-1973 experiment was to measure the decay spectrum of $K_2^{0-} \rightarrow \pi^+\pi^-\pi^0$. The design of the apparatus guaranteed that a significant number of $K_2^0 \rightarrow \pi^+\pi^-$ decays, those needed for a measurement of $|\eta_{+-}|$, would also be observed. With the announcement of Steinberger's result, however, the discrepancy between it and the previous measurements made another measurement important, and these events were used for that purpose.

The discrepancy between the old and new results has been commented on in the "Review of Particle Properties," the standard compilation of results in experimental particle physics:

There is a very large discrepancy between the old and new results for $|\eta_{+-}|$. ... The origin of the discrepancy ... is not known. ... We are troubled by this large, unexplained discrepancy. We feel that our normal procedure of averaging and increasing the error by a scale factor S to account for the discrepancy is not adequate for this case. The new results, when combined with the average of earlier results by that procedure give $(2.15 \pm 0.11) \times 10^{-3}$ $(S = 6.0)$. While this value and error makes some sense in that it nearly spans both incompatible sets of data we choose not to quote it. Instead, since the newer experiments are in principle superior (higher statistics, better acceptance, easier trigger conditions), we have chosen to average them separately from the earlier results.[20]

One is left to make one's own judgment.

Despite this obviously worrisome discrepancy, there has been no systematic study of the problem, let alone any resolution of it. This may be explained by the current lack of any theoretical prediction of the value of this parameter. It is the existence of the effect (i.e., $K_2^0 \rightarrow \pi^+\pi^-$) that is of importance, because it demonstrates violation of CP symmetry. Should such theoretical predictions be forthcoming, resolution of the problem will become more important and perhaps will even justify further remeasurements.

A recent analysis by Aronson and associates[21] of new measurements of $|\eta_{+-}|$ at high energies (30–110 GeV) has made resolution of the discrepancy between the two low energy values (~5 GeV) more important. Those authors found suggestive, although not convincing, evidence for an energy dependence of

$|\eta_{+-}|$ using only the high-energy measurements. The effect becomes more significant statistically if the low energy values are included. In fact, that group made two independent fits to the energy dependence of $|\eta_{+-}|$ using both values obtained at low energy. Clearly, the value of the energy dependence, if it exists, and its possible explanation depend on the correct value of $|\eta_{+-}|$ at low energy. This will no doubt encourage investigation of the causes of the discrepancy between the two sets of low-energy measurements, and perhaps lead to resolution of the problem. It may even lead to further low energy measurements of $|\eta_{+-}|$. A recent measurement was performed at an average K^0 momentum of 65 GeV/c by Coupal and associates.[22] That is quite close to the 70-GeV/c average momentum of the Aronson group. Coupal and associates obtained a value for $|\eta_{-+}|$ of $(2.28 \pm 0.06) \times 10^{-3}$, in good agreement with the average of the latest group of measurements $(2.27 \pm 0.022) \times 10^{-3}$ and in disagreement with the value of Aronson of $(2.09 \pm 0.02) \times 10^{-3}$. They concluded: "...we find no evidence in this experiment to support the existence of an energy dependence in η_{+-}."

History shows clearly the danger in jumping to even very well based conclusions. We have also discussed some of the circumstances under which repetitions seem reasonably justified, namely, (1) present or future utility on theoretical grounds of obtaining a more precise value, (2) the existence of previous discordant measurements of the quantity, or (3) when the measurement of the quantity comes as part of another measurement.

This episode raises the possibility of what one might call a bandwagon effect. Although each measurement was honestly made, they were, except for the first, conducted in the light of previous results. Obtaining a result for $|\eta_{+-}|$ from the "raw data" of the experiments is a complex process. It is certainly not obvious, without a very detailed investigation of the experiments and their analyses, how the values of $|\eta_{+-}|$ were obtained, or where the possible sources of discrepancy were. In any experiment, the sources of error, particularly systematic error, may be hidden and subtle. This is particularly true of the technically difficult experiments discussed earlier. The question of when to stop the search for sources of error is then very important. One psychologically plausible end point is when the result seems "right" (i.e., in agreement with previous results). This has been commented on in the "Review of Particle Properties," the standard reference for the physics community: "The old joke about the experimenter who fights the systematics until he or she gets the 'right' answer (read 'agrees with previous experiments') and then publishes contains a germ of truth. . . . A result can disagree with the average of all previous experiments by five standard deviations, and still be right."[23] A similar view was reported by Raymond Birge in his discussion of the evaluation of fundamental physical constants. He reported an explanation that he attributed to E. O. Lawrence:

In any highly precise experimental arrangement there are initially many instrumental difficulties that lead to numerical results far from the accepted value of the quantity being measured. It is, in fact, just such wide divergences that are the best indication of instrumental errors of one kind or another. Accordingly the experimenter searches for the source or sources of such errors, and continues to search until he gets a result close to the accepted value. *Then he stops!* [Emphasis in original.] But it is quite possible that he has still overlooked some source of error that was present also in previous work. In this way one can account for the close agreement of several different results and also for the possibility that all of them are in error by an unexpectedly large amount.[24]

It is the evidence presented in Figures 8.1 and 8.2, as well as belief in that attitude, which Peter Franken has called "intellectual phase locking,"[25] that has caused people to question the results of measurements of physical quantities and to suggest that scientists tend to underestimate the errors in such measurements.[26] A more sanguine view of the problem was given by Cohen and DuMond in their survey of the status of physical constants in 1965:[27] " . . . having done one's best with the available data, we must all learn not to be too surprised or disappointed if more highly developed methods subsequently reveal the presence of systematic errors unsuspected at the earlier date and of considerably larger magnitude than the earlier estimate of random error."[28] They, Birge, and Lawrence have all suggested that the safeguard against this problem is for such measurements to be performed by "different" methods. The histories of such measurements, particularly those for e, the charge on the electron, and c, the speed of light, often do show such major changes when a new, and presumably better, technique is used.[29] As noted in

Chapter 4, the two sets of measurements of $|\eta_{+-}|$ did use some what "different" experimental arrangements.

Contrary to the view expressed by Henrion and Fischoff,[30] who seem to want scientists to increase (and, in their view, better) their uncertainty estimates, so that such incidents will not occur, Cohen and DuMond argue against such a procedure. "It would be an equally grave mistake to recommend that the experimenter enlarge his error estimates" to take care of possible but unknown systematic errors. "Systematic errors in physical measurements do not obey any known statistics. There are, of course, limitations to the applicability of the experimental method. We simply have to learn the hard fact that, having arrived at a determination of a physical quantity and its estimated uncertainty in the light of all the best information available at a given epoch, this may prove at a later epoch, when we have more and better information to have been wrong."[31] Cohen and DuMond also argue against a single experimenter increasing his error estimate in order to be "safe":

The idea that an overestimate of error "for safety" is somehow more laudable or virtuous than an effort to be as accurate as one can with as little bias as possible in either magnifying or minimizing the standard deviation estimate, is somehow deplorably prevalent. We ask, for whom is such an overestimate "safe"? Certainly not for the general scientific community who wish to use the result. For them it is a concealment of the true facts regarding the results of the measurements. For the author of the result it may appear to be "safe" in the sense that at some later date his measurement might be less likely to be contradicted by later work; but even for him this unworthy timidity may be an illusory safety for, because of the unwarranted exaggeration of his error estimate, a crucial discrepancy, which might otherwise reveal some basically important new fundamental fact, may have been buried and lost forever as far as his reputation is concerned. For courageous men of science there is only one "safe" refuge, the plain, unvarnished truth as to their methods and results.[32]

What the history of these measurements shows is that usually there are many repetitions of the measurement of an important quantity, and large changes in the average values can and do sometimes occur. (For other examples, see Particle Data Group,[23] page S284, and Figure 8.2) These repetitions provide both oc-

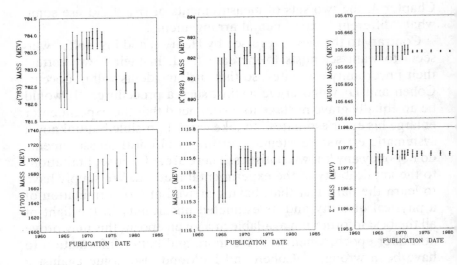

Figure 8.2. Some averages of elementary particle masses as a function of publication date.[23] Courtesy of Lawrence Berkeley Laboratory, University of California.

casional evidence for and examples of the checks on this possible bandwagon effect.

One may wonder, however, if such an anomalous experimental result that disagrees with the results of previous observations will fail to appear in the scientific literature, either because of self-censorship or because of a decision by journal referees and editors. Such a discrepancy may legitimately cause an experimenter to examine his work for possible sources of error, but it will not, in general, prevent submission of the results. We see this clearly in the case of Geweniger and associates[33] who submitted their anomalous measurement of $|\eta_{+-}|$ despite its 8-S.D. difference from the existing world average, and in the case of Fairbank and his collaborators, whose results on fractional charge were submitted despite the failure of all previous experiments to observe such charges, and in the face of skepticism by the physics community. The fact that both of those articles were published argues for openness on the part of journal referees and editors, a point supported by the histories of the measurements of various quantities given in the "Review of Particle Properties."

8.4   COOPER AND SEGRÈ: AN ACCUSATION OF
SCIENTIFIC FRAUD

In recent work, J. C. Cooper[34] accused Segrè and Chamberlain of fraud in the Nobel Prize-winning experiment in which they discovered the antiproton.[35] Cooper accused these physicists of failing to point out that their published data show evidence for faster-than-light particles, or tachyons. In addition, Cooper claimed that there had been a deliberate cover-up of this fact, not only by Segrè and Chamberlain but also by the entire physics community. Somewhat overdramatically, Cooper stated that, "The Segrè experiment is to the physics community what the Watergate tapes were to ex-President Nixon."[36]

Cooper's claims are completely without merit. First, there is no evidence in the original data of Segrè to support the conclusion that faster-than-light particles were observed. Second, despite Cooper's claim, there is nothing to indicate that the experimenters were aware of that conclusion and concealed it.

Cooper's method was to use the distance between two counters in the experiment and the measured time-of-flight distribution of pions between those two counters to determine the velocity of the pions. Using an assumed distance between the counters of forty feet and an average pion time of flight of 38.64 ± 0.73 nsec, he found[37] that the average pion velocity was $(1.05 \pm 0.02)c$.

Although Cooper made much of the published counter distance of forty feet, it is clear from comments by the experimenters[38] that that was only a crude measurement of the distance, and, as we shall see later, a reasonably accurate one. A much better estimate of the distance can be obtained by using the measured time of flight for the $1.19 - $ GeV/c antiprotons. The antiprotons have $\beta_{\bar{p}} = v/c = 0.785$. As pointed out in the Segrè paper, ionization losses caused by passage through the counters reduce $\beta_{\bar{p}}$ to 0.765 by the end of the beam line. A reasonable estimate for the average speed of the antiprotons is $\beta_{\bar{p}} = 0.775$. Thus the distance between the counters is $\ell = \beta_{\bar{p}} c t_{\bar{p}}$, where $t_{\bar{p}}$ is the average time of flight for the antiprotons. Using $t_{\bar{p}} = 51.12 \pm 1.40$ nsec (Cooper's calculations),[39] I obtain $\ell = 11.88 \pm 0.33$m or 38.98 ± 1.08 feet, in quite reasonable agreement with the estimate of forty feet.

Cooper claimed that any such change in the distance between the counters would destroy the published agreement between the antiproton and proton masses. That is incorrect. He assumed that the antiproton mass was determined by separately measuring the momentum and velocity of the antiprotons, and then calculating the antiproton mass. The published curve (Cooper,[34] Figure 4) was actually obtained by keeping the velocity selector constant and varying the beam momentum, using both protons and antiprotons. That resulted in relative measurements for the masses of the proton and antiproton, which gave the excellent agreement shown in the figure. It was, in fact, independent of the distance between the counters. It seems fair, then, to use the distance calculated earlier as the best estimate of the distance between the counters.

Using that distance 11.08 ± 0.33m and the measured average pion time of flight of 38.64 ± 0.73nsec, we obtain a pion velocity, $V_\pi = 3.075 \pm 0.103 \times 10^8$ m/sec $= (1.026 \pm 0.034)c$.[40] That is in quite reasonable agreement with the expected V_π for 1.19-GeV/c pions of $0.993c$.

Thus, there is no evidence to support Cooper's claim that faster-than-light particles were observed in that experiment.

Further evidence is provided by the experiment of Harrison and associates.[41] In that experiment, the beam momentum was found by comparing the calculated time-of-flight difference for π mesons and tritons (hydrogen nuclei of mass 3) as a function of momentum with the difference measured by two different methods. Similar calculations were performed for protons and deuterons and protons and π mesons. All three gave consistent values for the momentum of the beam. If the pions had been traveling at $1.05c$ (Cooper's value), the difference involving the time of flight of the pions would have been 5.9 nsec, which would have been easily observable (the time-of-flight measurements were done to better than ± 1 nsec accuracy) and would have destroyed the agreement among the three different methods for finding the momentum. It should also be emphasized that the experimental conditions were similar to those of the Segrè experiment. The two beam momenta were 1.19 GeV/c (Segrè) and 1.34 GeV/c, and the beam lengths were 40 and 108 feet.

His claim of a cover-up by the experimenters and by the physics community was based on his belief that his conclusion of faster-than-light particles was known to both, a fact nowhere in evidence.

Nevertheless, Cooper raised an important question: Do experimental results in disagreement with a well-corroborated and strongly believed theory tend to be suppressed in what we might call a theoretical bandwagon effect? Cooper attributed Segrè's alleged fraud to a desire not to contradict Einstein's special theory of relativity, which at the time did not appear to allow the existence of faster-than-light particles. There have indeed been searches for tachyons,[42] although, to be fair to Cooper, those searches occurred after a revival of interest in such particles following theoretical work in the 1960s that did allow tachyons.

More generally, however, given the importance of refutation or anomaly to the progress of science, a point emphasized by almost all modern philosophers of science, it would seem that results in disagreement with theory would be highly sought after, rather than suppressed. As Michael Kreisler, himself a searcher for tachyons, stated:

Every researcher hopes that his next experiment will yield a fundamental discovery – a discovery that will be considered not only essential to his particular sub-sub-specialty, but one that will radically change our view of the study of physics. Such experiments in the history of science are well-known; for example, the discovery of the pion or Yukawa particle or the discovery of superconductivity. The desire to participate in such a discovery is partially responsible for the large number of experiments testing the predictions of fundamental laws, some of which have been found to fail to hold when tested extremely carefully. It is a healthy sign for the study of physics that there are no "sacred cows"; if there is any reasonable chance that something new exists, researchers will spare no effort in the search. The quest for particles travelling faster than the speed of light, testing our understanding of special relativity, is one such search.[43]

Two of the foregoing case studies involved results in disagreement with a strongly believed theory – the results on electron scattering, which disagreed with Dirac's electron theory, and those showing CP violation, a strongly held symmetry principle – thus arguing against any suppression of this kind of result.

8.5   DISCUSSION

The case histories discussed earlier seem to indicate that the normal procedures of the scientific community do provide adequate safeguards against fraud. The reason for this is the repetition of experiments, particularly those with theoretical importance.[44] The two cases of fraud in physics discussed earlier were adequately handled by repetition, and repetition also guards against the possibility of an experimental bandwagon effect.

A legitimate question may be raised whether or not detailed examination of four episodes, involving only the history of twentieth-century physics, provides sufficient support for the generalization that the normal procedures of science constitute sufficient safeguards against fraud. I suggest that it does. The episodes cover a period of over sixty years, and several subspecialties in physics involving different groups of scientists. In addition, I believe that experiments are more clearly defined in physics than in the social sciences, in which questions concerning controls, relevant data, and replications are not easy to answer.

Schmaus[45] has correctly suggested that the procedures of the scientific community are designed to deal with error, not with fraud. What should be emphasized, however, is that from the point of view of science, incorrect fraudulent results and error are indistinguishable. If fraudulent results happen to be correct (e.g., Millikan's trimming discussed earlier), no problem arises for science. The case is, of course, different when deciding which researchers should be funded. Here one might wish to exclude all those guilty of fraud. A more serious problem occurs when social policy may be based on such results, but that discussion is beyond the scope of this book.

It should also be emphasized that the more important a result is, either because of its relation to existing theories or because of its call for new theories, the more likely it is to be repeated.[46] This would seem to pose a problem for the prospective defrauder. The argument proceeds as follows: One is more tempted to fraud the more important a result is, because then one will acquire more prestige, money, and so forth. The more important the result is, the more likely the fraud is to be uncovered, because of the greater probability of repetition. Thus, the more important the fraud, the more likely it is that one will be severely punished, because discovered fraud is severely punished. It is therefore counterproductive to engage in fraud on important results. Fraud on unimportant results seems hardly worthwhile, and is less likely to occur. It is also less likely to be detected, but that does not seem to pose any great danger to science.

In the cases of fraud mentioned earlier,[1] Soman, Long, Darsee, and Gullis lost their jobs, and although Felig has returned to a tenured professorship at Yale, he was forced to resign a prestigious chair at Columbia. At the very least, each person's scientific career has been blighted. It is this severe punishment that shows that the scientific community still holds strongly to the ideal of "disinterestedness." Cooper apocalyptically stated that, "The name 'Segrè' will mean to science what the name 'Judas Iscariot' means to Christianity".[47] Although his accusation of Segrè was false, such a comparison would be apt for real fraud. Fraud in science is a betrayal of science, and being found guilty of fraud removes one from the community of scientists.

One may well ask why, if the penalties are so severe, did the scientists involved in the recent biomedical episodes engage in fraud. One may speculate that in each case the scientists believed that they were in fact correct and that they were only manufacturing the data until real data became available.[48]

Despite the strong evidence presented earlier that fraud is not a major problem for science, some small worries remain. As Francis Bacon stated, "And therefore it was a good answer that was made by one who, when they showed him hanging in a temple a picture of those who had paid their vows as having escaped shipwreck, and would have him say whether he did not now acknowledge the power of the gods – 'Aye,' asked he again, 'but where are they painted that were drowned after their vows?' "[49] Where, indeed, are the frauds that have not been uncovered?

Conclusion

I would hope that after reading this study the reader will have an increased appreciation of the varied and important roles that experiment plays in physics. The reader should conclude that experiment has a philosophically legitimate role in the choice between competing theories and in confirmation of theories or hypotheses and that there are good reasons for believing in experimental results. These are rather commonplace and obvious conclusions. Why should one write or, perhaps more important, read a book that concludes so little? I believe that detailed historical study of actual, as opposed to mythical, experiments is worthwhile. Some philosophers and sociologists of science have sought to deny, or to minimize, the role that experiment plays in theory choice and to cast doubt on the validity of experimental results. This book is my attempt to answer them.

Notes

INTRODUCTION

1. A rather sketchy survey among my colleagues in the Physics Department at the University of Colorado indicated that very few of them knew of the work of Lummer and Pringsheim.
2. O. Lummer and E. Pringsheim, "Die Verteilung der Energie im Spectrum des schwarzen Körpers," *Verh. d. D. Phys. Ges.*, *1* (1899), 23–41; O. Lummer and E. Pringsheim, "Die Verteilung der Energie im Spectrum des schwarzen Körpers und des blanken Platins," ibid., *1* (1899), 215–35; O. Lummer and E. Pringsheim, "Ueber die strahlung des schwarzen Körpers für langen Wellen," ibid., *2* (1900) 163–80; O. Lummer and P. R. E. Jahnke, "Ueber die Spectralgleichung des schwarzen Körpers und des blanken Platins," *Annalen der Physik*, *3* (1900), 283–97.
3. H. Rubens and F. Kurlbaum, "Ueber die Emission langwelliger Wärmestrahlen durch den schwarzen Körper bei verschiedenen Temperaturen," *Berl. Ber.* (1900), 929–41.
4. T. S. Kuhn, *Black-Body Theory and the Quantum Discontinuity, 1894–1912* (Oxford University Press, 1978).
5. L. Cooper, *Aristotle, Galileo and the Tower of Pisa* (Ithaca: Cornell University Press, 1935).
6. C. G. Adler and B. Coulter, "Galileo and the Tower of Pisa Experiment," *AJP*, *46* (1978), 199–201.
7. A. Franklin, "Galileo and the Leaning Tower: An Aristotelian Interpretation," *Physics Education*, *14* (1979), 60–3.
8. J. Worrall, "Thomas Young and the 'Refutation' of Newtonian Optics: A Case-study in the Interaction of Philosophy of Science and History of Science," in C. Howson (ed.), *Method and Appraisal in the Physical Sciences* (Cambridge University Press, 1976), 107–79.
9. See, for example, the account given in I. Lakatos, "Falsification and the Methodology of Scientific Research Programmes," in I. Lakatos and A. Musgrave (eds.), *Criticism and Growth of Knowledge* (Cambridge University Press, 1970), 91–196. Lakatos reported that Michelson performed a long series of experiments from 1881 to 1935 (p. 163). As Ian Hacking (note 12, p. 260) noted, "The experiments he did from 1931 to 1935 must be on the astral plane, for he died in 1931." Hacking presented a much more reasonable view.

10. L. Swenson, "The Michelson-Morley-Miller Experiments Before and After 1905," *Journal for the History of Astronomy*, *1* (1970), 56–78.

11. M. Handschy, "Re-examination of the 1887 Michelson-Morley Experiment," *AJP*, *50* (1982), 987–90.

12. I. Hacking, *Representing and Intervening* (Cambridge University Press, 1983).

13. P. Galison, "Theoretical Predispositions in Experimental Physics. Einstein and the Gyromagnetic Experiments, 1915–25," *Historical Studies in the Physical Sciences*, *12(2)* (1982), 285–323; P. Galison, "How the First Neutral Current Experiments Ended," *Reviews of Modern Physics*, *55* (1983), 477–509; A. Pickering, "The Hunting of the Quark," *Isis*, *72* (1981), 216–36; B. Wheaton, *The Tiger and the Shark* (Cambridge University Press, 1983); R. Stuewer, *The Compton Effect* (New York: Science History Publications, 1975). Philosophical discussion is a major part of Worral (note 8) and J. Worrall, "The Pressure of Light: The Strange Case of the Vacillating Crucial Experiment," *Studies in History and Philosophy of Science*, *13(2)* (1982), 133–71, as well as D. Gooding, "Empiricism in Practice: Teleology, Economy, and Observation in Faraday's Physics," *Isis*, *73* (1982), 46–67.

14. K. R. Popper, *The Logic of Scientific Discovery* (New York: Harper & Row, 1968); C. Glymour, *Theory and Evidence* (Princeton University Press, 1980); Hacking (note 12); D. Shapere, "The Concept of Observation in Science and Philosophy," *Philosophy of Science*, *49* (1982), 485–525.

15. P. Feyerabend, *Against Method* (Atlantic Highlands, London: Humanities Press, 1975).

16. T. S. Kuhn, *The Structure of Scientific Revolutions* (University of Chicago Press, 1970).

17. Kuhn (note 16, p. 148).

18. Lakatos (note 9, p. 119).

19. An interesting discussion of the entire problem, including the original essays of Duhem and Quine, is contained in Sandra Harding (ed.), *Can Theories Be Refuted?* (Dordrecht: D. Reidel, 1976).

20. W. Quine, *From a Logical Point of View* (Cambridge: Harvard University Press, 1953), 43.

21. J. Earman (ed.), *Testing Scientific Theories* (Minneapolis: University of Minnesota Press, 1983); Harding (note 19); P. Achinstein, *The Concept of Evidence* (Oxford University Press, 1983).

22. Glymour (note 14) and Shapere (note 14).

CHAPTER 1

1. T. D. Lee and C. N. Yang, "Question of Parity Nonconservation in Weak Interactions," *PR*, *104* (1956), 254–8. A very nice history of this episode appears in B. Maglic (ed.), *Adventures in Experimental Physics*, gamma volume (Princeton: World Science Education, 1973), 93–162.

2. Until recently, physicists generally divided the interactions or forces between bodies into four classes: (1) the strong or nuclear interaction, which

is responsible for holding the atomic nucleus together, (2) the electromagnetic interaction, which describes the forces between charged objects, (3) the weak interactions, which are responsible for nuclear beta decay and some of the decays of elementary particles, (4) the gravitational interaction. The order of the relative strengths of these interactions is as follows: strong : electromagnetic : weak : gravitational $= 1 : 10^{-2} : 10^{-14} : 10^{-38}$.

3. C. S. Wu, E. Ambler, R. W. Hayward, D. D. Hoppes, and R. P. Hudson, "Experimental Test of Parity Nonconservation in Beta Decay," *PR, 105* (1957), 1413–15.

4. Richard L. Garwin, Leon M. Lederman, and Marcel Weinrich, "Observation of the Failure of Conservation of Parity and Charge Conjugation in Meson Decays: The Magnetic Moment of the Free Muon," *PR, 105* (1957), 1415–17.

5. Jerome L. Friedman and V. L. Telegdi, "Nuclear Emulsion Evidence for Parity Nonconservation in the Decay Chain π^+-μ^+-e^+", *PR, 105* (1957), 1681–2.

6. For a detailed discussion of the concept of parity, see, for example, W. M. Gibson and B. R. Pollard, *Symmetry Principles in Elementary Particle Physics* (Cambridge University Press, 1976), Chapter 5; or Hans Frauenfelder and Ernest M. Henley, *Nuclear and Particle Physics* (Reading, Mass.: W. A. Benjamin, 1975), Chapter 4.

7. The careful reader will note that the term "parity" has been used in two ways. The first is as the operation of spatial inversion, and the second is as a property or numerical quantity associated with the system. When we speak of the conservation of parity, we shall normally be referring to the second usage.

8. See, for example, George Gamow and John M. Cleveland, *Physics, Foundations and Frontiers* (Englewood Cliffs, N.J.: Prentice-Hall, 1969), 220–2.

9. A dextro-rotary isomer rotates it to the left.

10. Gibson and Pollard (note 6, p. 119).

11. Recent work by S. F. Mason and G. E. Tranter, "The Electroweak Origin of Biomolecular Handedness," *Proc. Roy. Soc. (London), A397* (1985), 45–65, using the contemporary theory of electroweak interactions, shows that for certain molecules, the L form has a very slightly lower energy than the D form. They suggest that this accounts for the dominance of L-form molecules and that biomolecular handedness was not a matter of chance. A less technical discussion, including the history of nineteenth-century chemistry work on asymmetric molecules, is Stephen Mason, "Biomolecular Handedness," *Chemistry in Britain, 21* (1985), 538–45. See also Stephen F. Mason, "Origins of Biomolecular Handedness," *Nature, 311* (1984), 19–23.

12. Hans A. Bethe and Phillip Morrison, *Elementary Nuclear Theory*, second ed. (New York: Wiley, 1956), 152–3; Gibson and Pollard (note 6, p. 133).

13. Bethe and Morrison (note 12, p. 153).

14. O. Laporte, "Die Struktur des Eisenspektrums," *Z. Phys., 23* (1924), 135–75.

15. E. Wigner, "Einige Folgerungen aus der Schrödingerschen Theorie für die Termstrukturen," *Z. Phys., 43* (1927), 624–52.

16. A very amusing early reference to this occurs in a paper by P. Jordan and R. de L. Kronig, "Movement of the Lower Jaw of Cattle during Mastication," *Nature, 120* (1927), 807. In that paper, Jordan and Kronig noted that the chewing motion of cows is not straight up and down, but rather is either a left-circular or right-circular motion. They reported on a survey of cows in Sjaelland, Denmark, and observed that 55% were right-circular and 45% left-circular, a ratio they regarded as consistent with unity.

17. Frauenfelder and Henley (note 6, p. 359).

18. There are two papers that discussed the logical possibility of parity non-conservation prior to 1956, and we shall discuss these in detail later: E. M. Purcell and N. F. Ramsey, "On the Possibility of Electric Dipole Moments for Elementary Particles and Nuclei," *PR, 78* (1950), 807; G. C. Wick, A. S. Wightman, and E. Wigner, "The Intrinsic Parity of Elementary Particles," *PR, 88* (1952), 101–5.

19. We note, again, that in a state of orbital angular momentum l, the spatial parity of the wave function is given by $(-1)^l$. See, for example, Gibson and Pollard (note 6, p. 120).

20. C. N. Yang, "Present Knowledge About the New Particles," *Reviews of Modern Physics, 29* (1957), 231–5.

21. R. H. Dalitz, "On the Analysis of τ-Meson Data and the Nature of the τ-Meson," *Phil. Mag., 44* (1953), 1068–80; "Decay of τ Mesons of Known Charge," *PR, 94* (1954), 1046–51.

22. T. D. Lee, "History of Weak Interactions," speech at Columbia University Physics Department, March 26, 1971, unpublished, p. 10.

23. See discussion in *High Energy Nuclear Physics*, proceeding of the sixth annual Rochester conference, 3–7 April 1956, compiled and edited by J. Ballam, V. L. Fitch, T. Fulton, K. Huang, R.R. Rau, and S.B. Trieman (New York: Interscience, 1957), Chapter V, pp. 29–31, and Chapter VIII, pp. 1–28; and Yang (note 20).

24. Lee (note 22, p. 11).

25. Lee and Yang (note 1); Lee (note 22).

26. K. Siegbahn (ed.), *Beta- and Gamma-Ray Spectroscopy* (Amsterdam: North Holland, 1955).

27. A pseudo-scalar is a quantity that changes sign under the parity operation.

28. Lee (note 22, p. 12).

29. R. T. Cox, C. G. McIlwraith, and B. Kurrelmeyer, "Apparent Evidence of Polarization in a Beam of β-Rays," *Proceedings of the National Academy of Sciences (USA), 14* (1928), 544–9.

30. Carl T. Chase, "The Scattering of Fast Electrons by Metals. II. Polarization by Double Scattering at Right Angles," *PR, 36* (1930), 1060–5.

31. In a private conversation with the author, T. D. Lee stated that he and Yang did not become aware of these experiments until long after their initial work.

32. C. N. Yang, "Nobel Prize Address," in *Nobel Lectures, Physics, 1942–1962* (Amsterdam: Elsevier, 1964), 398; Maglic (note 1, p. 98).

33. Lee and Yang (note 1).

34. Lee and Yang (note 1, p. 254).

35. Lee and Yang (note 1, p. 254).

36. Note 18.
37. Lee and Yang (note 1, p. 254).
38. Lee and Yang (note 1, p. 255).
39. Lee and Yang (note 1, p. 255).
40. Wu et al. (note 3).
41. Wu et al. (note 3, p. 1413).
42. Lee and Yang (note 1, p. 257).
43. Garwin et al. (note 4).
44. Friedman and Telegdi (note 5).
45. Garwin et al. (note 4, p. 1416).
46. Garwin et al. (note 4, p. 1415). The authors acknowledged that they knew of the preliminary results of Wu's experiment before they actually began their own work. For a day-by-day (sometimes almost hour-by-hour) account of this, see Maglic (note 1) and also Lee (note 22).
47. Friedman and Telegdi, (note 5).
48. Friedman and Telegdi (note 5, p. 1681).
49. T. D. Lee, R. Oehme, and C. N. Yang, "Remarks on Possible Noninvariance under Time Reversal and Charge Conjugation," *PR, 106* (1957), 340–5.
50. Although both the principle of parity conservation (P) and the principle charge-conjugation invariance (C) had been shown to be violated, physicists believed that the product of the two operations, CP, was still conserved. This remained true until 1964, when Christenson, Cronin, Fitch, and Turlay showed that this invariance principle was violated. This will be discussed in Chapter 3.
51. Besides those experiments discussed in the text, some of the other experiments that showed parity violation are as follows: the $\pi \to \mu \to$ e decay results were confirmed by yet another technique, that of liquid-hydrogen bubble chambers, by A. Abashian et al., "Angular Distribution of Positrons Observed in a Liquid Hydrogen Bubble Chamber," *PR, 105* (1957), 1927–8. Circular polarization of γ rays emitted after β decay was shown by H. Schopper, "Circular Polarization of γ-rays: Further Proof for Parity Failure in β Decay," *Phil. Mag., 2* (1957), 710–13, and by F. Boehm and A. H. Wapstra, "Beta-Gamma Circular Polarization Correlation Measurements," *PR, 106* (1957), 1364–6. Circular polarization of bremsstrahlung (γ rays) emitted by β rays was demonstrated by M. Goldhaber, L. Grodzins, and A. Sunyar, "Evidence for Circular Polarization of Bremsstrahlung Produced by Beta Rays," *PR, 106* (1957), 826–8. Lorne A. Page and Milton Heinberg, "Measurement of the Longitudinal Polarization of Positrons Emitted by Sodium–22," *PR, 106* (1957), 1220–4, and S. S. Hanna and R. S. Preston, "Positron Polarization Demonstrated by Annihilation in Magnetized Iron," *PR, 106* (1957), 1363–4, measured the longitudinal polarization of positrons (positive electrons) emitted in radioactive decay. This is only a partial list, but it does show the amount of interest generated in the subject. Wu and her collaborators published further results in E. Ambler et al., "Further Experiments on β Decay of Polarized Nuclei," *PR, 106* (1957), 1361–6, as did Friedman and Telegdi,

250 Notes to pages 20–27

"Nuclear Emulsion Evidence for Parity Nonconservation in the Decay Chain $\pi^+ \to \mu^+ \to e^+$," *PR*, *106* (1957), 1290–3.

52. Notes 29 and 30.
53. T. D. Lee and C. N. Yang, "Parity Nonconservation and a Two-Component Theory of the Neutrino," *PR*, *105* (1957), 1671–4.
54. H. Frauenfelder et al., "Parity and the Polarization of Electrons from Co^{60}," *PR*, *106* (1957), 386–7.
55. Frauenfelder et al. (note 54, p. 386).
56. P. E. Cavanagh et al., "On the Longitudinal Polarization of β-particles," *Phil. Mag.*, 2 (1957), 1105–12.
57. A. De Shalit, S. Cuperman, H. J. Lipkin, and T. Rothem, "Detection of Electron Polarization by Double Scattering," *PR*, *107* (1957), 1459–60; "Measurement of Beta-Ray Polarization of Au^{198} by Double Coulomb Scattering," *PR*, *109* (1958), 223–4.
58. O. B. Klein "Nobel Prize Presentation Speech," in *Nobel Lectures, Physics, 1942–1962* (Amsterdam: Elsevier, 1964), 390.
59. C. S. Wu, in Maglic (ed.) (note 1, p. 104).
60. V. L. Telegdi, in Maglic (ed.) (note 1, p. 132).
61. See note 16 for an earlier lighthearted view of parity conservation.
62. H. Weyl, "Elektron und Gravitation. I," *Z. Phys.*, 56 (1929), 330–52. W. Pauli, "Die allgemeinen Prinzipien der Wellenmechanik," *Handbuch der Physik, Ser. 2*, 24 (1933), 83–272.
63. Pauli (note 62, p. 226).
64. Lee and Yang (note 53).
65. L. Landau, "On the Conservation Laws for Weak Interactions," *Nuclear Physics*, 3 (1957), 127–31.
66. A. Salam, "On Parity Conservation and Neutrino Mass," *NC*, 5 (1957), 299–301.
67. W. Pauli, quoted in Jeremy Bernstein, *A Comprehensible World* (New York: Random House, 1967), 59.
68. W. Pauli, in Maglic (ed.) (note 1, p. 122).
69. Frauenfelder and Henley (note 6, p. 389). Feynman himself repeated the same story in his autobiographical *Surely You're Joking, Mr. Feynman* (New York: Norton, 1985), 248.
70. T. D. Lee, private communication.
71. David Halliday, *Introductory Nuclear Physics* (New York: Wiley, 1955), 38.
72. I. I. Rabi, quoted in Bernstein (note 67, p. 58).
73. W. Pauli, quoted in Bernstein (note 67, p. 60).
74. K. K. Darrow, quoted in Bernstein (note 67, p. 61).
75. Daniel Sullivan, D. Hywel White, and Edward J. Barboni, "The State of Science: Indicators in the Specialty of Weak Interactions," *Social Studies of Science*, 7 (1977), 167–200.
76. Sullivan et al. (note 75, p. 180).
77. Sullivan et al. (note 75, pp. 186–7).
78. Peter J. Brancazio, *The Nature of Physics* (New York: Macmillan, 1975), 684. To see the different treatments of a topic in two editions of the same textbook, one before and one after 1957, see Leonard Schiff, *Quantum*

Mechanics (New York: McGraw-Hill, second ed. 1955, p. 160, third ed. 1968, pp. 253–4).

79. V. L. Telegdi, in Maglic (ed.) (note 1, pp. 134–5).
80. The footnote quoted in the text appears in Friedman and Telegdi (note 5, p. 1682).
81. C. Baltay et al., "Experimental Evidence Concerning Charge-Conjugation Noninvariance in the Decay of the η Meson," *PRL, 16* (1966), 1224–8. The term "standard deviation" is a measure of the precision of an experimental result. Assuming that the error analysis has been done correctly, the probability that the correct answer is within one standard deviation of the stated result is about 68%, and within two standard deviations, 95%.
82. S. Goudsmit, in Maglic (ed.) (note 1, p.137).
83. R. Garwin, in Maglic (ed.) (note 1, p. 137); also quoted in Lee (note 22, pp. 16–17).
84. Lee (note 22, p. 15).
85. Purcell and Ramsey (note 18).
86. Purcell and Ramsey (note 18, p. 807).
87. Lee and Yang (note 1, p. 255).
88. See N. F. Ramsey, *Molecular Beams* (Oxford University Press, 1956), 202.
89. T. D. Lee, private communication.
90. Wick et al. (note 18).
91. Wick et al. (note 18, p. 101).
92. Wick et al. (note 18, p. 104).
93. G. D. Rochester and C. C. Butler, "Evidence for the Existence of New Unstable Elementary Particles," *Nature, 160* (1947), 855–7.
94. A. M. Shapiro, "Table of Properties of the 'Elementary' Particles," *Reviews of Modern Physics, 28* (1956), 165–70.
95. M. Shapiro, "Mesons and Hyperons," *AJP, 4* (1956), 196–204.
96. C. N. Yang, in Ballam et al. (eds.) (note 23, Chapter VIII, p. 2).
97. A. Pais, "Some Remarks on the V-particles," *PR, 86* (1952), 663–72.
98. M. Gell-Mann, "Isotopic Spin and New Unstable Particles," *PR, 92* (1953), 833–4; T. Nakano and K. Nishijima, "Charge Independence for V-particles," *Progress in Theoretical Physics, 10* (1953), 581–2; K. Nishijima, "Some Remarks on the Even-Odd Rule," ibid., *12* (1954), 107–8; K. Nishijima, "Charge Independence Theory of V Particles," ibid., *13* (1955), 285–304; M. Gell-Mann, "The Interpretation of the New Particles as Displaced Charge Multiplets," *NC Suppl., 4* (1956), 848–66.
99. C. N. Yang, in Ballam et al. (eds.) (note 23, Chapter VIII, p. 3).
100. C. N. Yang, in Ballam et al. (eds.) (note 23, Chapter VIII, p. 4).
101. Sullivan et al. (note 75, pp. 178–80).
102. Yang (note 20, p. 231).
103. T. D. Lee and J. Orear, "Speculations on Heavy Mesons," *PR, 100* (1955), 932–3.
104. L. Alvarez, in Ballam et al. (eds.) (note 23, Chapter V, pp. 28–30).
105. T. D. Lee and C. N. Yang, "Mass Degeneracy of the Heavy Mesons," *PR, 102* (1956), 290–1.
106. R. Marshak, in Ballam et al. (eds.) (note 23, Chapter VIII, p. 18).

107. J. Orear, G. Harris, and S. Taylor, "Spin and Parity Analysis of Bevatron τ Mesons," *PR, 102* (1956), 1676–84.
108. C. N. Yang, in Ballam et al. (eds.) (note 23, Chapter VIII, pp. 27–8).
109. This work was done in collaboration with Professor Howard Smokler. I am grateful to him for allowing me to use it here.
110. See, for example, the essays by Duhem and Quine in S. Harding (ed.), *Can Theories Be Refuted?* (Dordrecht: D. Reidel, 1976), 1–64. Although Duhem and Quine argued against the possibility of any single experiment being crucial, their arguments were also directed against the possibility of any set of experiments being crucial. We are concerned here with the latter case, because that is, in fact, what occurred.
111. This statement is strictly true only for the leptonic and semileptonic weak interactions, which involve neutrinos. There are, in fact, some nonleptonic weak decay amplitudes that do conserve parity. There were several papers that applied parity-conserving schemes to these processes. It is important to note, however, that none of these papers cast any doubt on parity nonconservation in general. An amusing and interesting feature of several of these papers is that they dealt with the θ-τ puzzle, which was exactly the problem that stimulated Lee and Yang to suggest parity nonconservation. S. B. Treiman and H. H. Wyld [*PR, 106* (1957), 1320–3] proposed a parity-mixing scheme to explain θ and τ decays. However, that scheme was experimentally ruled out by F. Eisler, R. Plano, N. Samios, M. Schwartz, and J. Steinberger [*PR, 107* (1957), 324–5]. R. Gatto [*Nuclear Physics, 5* (1958), 235–55] also attempted to solve the problem without recourse to parity nonconservation. He stated, however, that "Solutions of the τ–θ problem are sought whereby one could avoid the conclusion that parity is violated also in weak interactions not involving neutrinos. It is shown that no such solutions can be constructed that are theoretically acceptable, and experimental tests are indicated for a definite disproof of such models." W. H. Nichols [*Annals of Physics, 5* (1958), 141–55] also proposed a parity-conserving scheme for K–π decay: "Now that this conservation law [parity] has been shown not to hold in beta decay, pion decay, muon decay, and K → μ + ν (all involving neutrinos), and Λ → π + p (involving no neutrinos), one ought to suspect that non-neutrino modes of K decay do not conserve parity either. However, since it is impossible experimentally to detect parity nonconservation in the decay of a spin-zero K meson, and since such an investigation may be significant for the decay of other bosons, it will be worthwhile to investigate theoretically one more parity conserving K–π interaction in hopes that it may fit the K–π data."
112. Such clarity avoids the difficulty posed by an anomalous result for a theory that is strongly believed on the basis of other experimental evidence. An example of this might be the failure to observe the forward–backward asymmetry in the double scattering of electrons predicted by Mott, on the basis of Dirac's theory of the electron, which was strongly supported by spectrosopic evidence. It would be interesting to examine this problem in the light of recent successful attempts to unify the weak and electromag-

netic interactions, as well as more general unifying theories, but that is left for further study.
113. I am grateful to Jon Dorling and H. R. Post for pointing this out to me.

CHAPTER 2

1. R. T. Cox, C. G. McIlwraith, and B. Kurrelmeyer, "Apparent Evidence of Polarization in a Beam of β-Rays," *Proceedings of the National Academy of Sciences (USA), 14* (1928), 544–9; Carl T. Chase, "The Scattering of Fast Electrons by Metals. II. Polarization by Double Scattering at Right Angles," *PR, 36* (1930), 1060–5.
2. L. Grodzins, "The History of Double Scattering of Electrons and Evidence for the Polarization of Beta Rays," *Proceedings of the National Academy of Sciences (USA), 45* (1959), 399–405, contains an excellent discussion.
3. Professor Arthur Roberts, a graduate student at New York University in the mid-1930s, where these experiments were performed, recalls some discussions of parity in connection with these results (private communication). Neither Cox nor Kurrelmeyer recalls any such discussions. Professor Max Dresden remembers a lecture given in 1939 by Otto Laporte that mentioned parity violation in connection with the Cox and Chase experiments. No mention of this appears in the literature.
4. L. de Broglie, "Ondes et Quanta," *CR, 177* (1923), 507–10; "Quanta de Lumiere, Diffraction et Interferences," *CF, 177* (1923), 548–50; "Sur la Definition Générale de la Correspondance entre Onde et Mouvement," *CR, 179* (1924), 39–40; "A Tentative Theory of Light Quanta," *Phil. Mag., 47* (1924), 446–8.
5. C. Davisson and L. H. Germer, "Diffraction of Electrons by a Crystal of Nickel," *PR, 30* (1927), 705–40.
6. G. E. Uhlenbeck and S. Goudsmit, "Spinning Electrons and the Structure of Spectra," *Nature, 117* (1926), 264–5.
7. C. G. Darwin, "The Electron as a Vector Wave," *Proc. Roy. Soc. (London), A116* (1927), 227–53.
8. Cox et al. (note 1, p. 544).
9. N. F. Mott, "Scattering of Fast Electrons by Atomic Nuclei," *Proc. Roy. Soc. (London), A124* (1929), 425–42; "The Polarisation of Electrons by Double Scattering," *Proc. Roy. Soc. (London), A135* (1932), 429–58.
10. Grodzins (note 2, pp. 399–400).
11. Grodzins (note 2, pp. 400–1).
12. Cox et al. (note 1, p. 545).
13. Cox et al. (note 1, p. 546).
14. Cox et al. (note 1, p. 547).
15. Cox et al. (note 1, p. 548).
16. According to modern theory and experiment, the polarization of the electrons is proportional to v/c, where v is the electron velocity and c is the speed of light.
17. Cox et al. (note 1, p. 548).

18. C. Chase, "A Test for Polarization in a Beam of Electrons by Scattering," *PR*, *34* (1929), 1069–74.
19. Chase (note 18, p. 1071).
20. Chase (note 18, p. 1073).
21. Carl T. Chase, "The Scattering of Fast Electrons by Metals. I. The Sensitivity of the Geiger Point-Discharge Counter," *PR*, *36* (1930), 984–7.
22. Chase (note 21, p. 987).
23. Chase (note 1).
24. Chase (note 1, p. 1064).
25. Cox et al. (note 1, p. 547).
26. Cox et al. (note 1, p. 547).
27. Cox et al. (note 1, p. 548).
28. Cox et al. (note 1, p. 548).
29. See Grodzins (note 2) for some valuable discussion.
30. "Plural scattering" is the term applied to the large-angle scattering that results from a few, as opposed to many, smaller-angle scatters.
31. Grodzins (note 2, p. 401).
32. L. Grodzins, in B. Maglic (ed.), *Adventures in Experimental Physics*, gamma volume (Princeton: World Science Education, 1973), 154–60.
33. Grodzins (note 32, p. 160).
34. R. T. Cox, private communication. In a subsequent letter, Cox has informed me that he is now convinced by my analysis that he had done a correct experiment but had made a slip in the coordinate systems.
35. R. T. Cox, in Maglic (ed.) (note 32, p. 149).
36. Wigner's paper on parity conservation was not published until 1927. In fact, the term "parity" was not even in use at that time. I do not know precisely when the term made its first appearance.
37. B. Kurrelmeyer, private communication.
38. R. T. Cox, private communication.
39. Cox, in Maglic (ed.) (note 32, p. 149). Cox points out that this statement was in fact an editorial emendation by B. Maglic and seems to convey an unreasonable reproach to the physics community, which was not Cox's intent (private communication). Inclusion of the next sentence gives a better statement of his views: "Only by viewing it from the new theoretical framework and experimental observations of the late 50's could our results be comprehended."
40. As discussed earlier, electrons that are initially unpolarized, such as those from thermionic sources, can give rise only to 0°–180° asymmetry. In order to get 90°–270° asymmetry, the electrons must have an initial longitudinal polarization, as the electrons from β decay possess.
41. H. Frauenfelder and E. Henley, *Nuclear and Particle Physics* (Reading, Mass.: W. A. Benjamin, 1975), 392.
42. R. T. Cox, private communication.
43. Cox, in Maglic (ed.) (note 32, p. 149).
44. E. Chargaff, *Science*, *172* (1971), 637.
45. S. Goudsmit, "It Might as Well be Spin," *Physics Today*, *29(6)* (1976), 40–3; quotation from p. 42.

46. H. A. Tolhoek, "Electron Polarization, Theory and Experiment," *Reviews of Modern Physics*, 28 (1956), 277–98.

47. F. Wolf, "Versuch über die Polarisations fahigkeit eines Elektronenstrahls," *Z. Phys.*, 52 (1928), 314–17.

48. Bohr's argument appears in Mott (note 9 [1929]).

49. E. Rupp, "Versuche zur Frage nach einer Polarisation der Elektronenwelle," *Z. Phys.*, 53 (1929), 548–52.

50. A. F. Joffé and A. N. Arsenieva, "Expériences sur la Polarisation des Ondes Electroniques," *CR*, 188 (1929), 152–3.

51. C. J. Davisson and L. H. Germer, "An Attempt to Polarise Electron Waves by Reflection," *Nature*, 122 (1928), 809.

52. J. Frenkel, "Sur l'impossibilité de Polariser des Ondes Cathodiques par Reflexion," *CR*, 188 (1929), 153–5.

53. C. J. Davisson and L. H. Germer, "A Test for Polarization of Electron Waves by Reflection," *PR*, 33 (1929), 760–72.

54. Davisson and Germer (note 53, pp. 771–2).

55. Mott (note 9 [1929]).

56. P. A. M. Dirac, "The Quantum Theory of Electrons," *Proc. Roy. Soc. (London)*, A117 (1928), 610–24. Dirac was awarded the Nobel Prize in 1933 for this work.

57. N. F. Mott, "Polarization of a Beam of Electrons by Scattering," *Nature*, 128 (1931), 454.

58. Mott (note 9 [1932]).

59. Mott (note 9 [1929]).

60. Sir Nevill Mott remarked (private communication): "The extraordinary thing is that I had no idea that early attempts, unsuccessful, to find the effect I predicted were due to the nature of the β-ray source. I suppose that when I went to Bristol in 1933 I got so absorbed in solid state that I have paid very little attention to this sort of thing. Also I am not very proud of that particular paper. . . . I made some numerical mistakes in the first version which had to be corrected. However, I remember that Pauli read it, liked it, and offered me a place in his group at Zurich which regrettably I could not take up, having already accepted something else."

61. Chase (note 18, p. 1073).

62. Chase (note 1, p. 1064).

63. E. Rupp, "Ueber eine unsymmetrische Winkelverteilung zweifach reflektierter Elektronen," *Z. Phys.*, 61 (1930), 158–69.

64. Chase (note 1).

65. E. Rupp, "Ueber eine unsymmetrische Winkelverteilung zweifach reflektieter Elektronen," *Natur.*, 18 (1930), 207.

66. E. Fues and H. Hellmann, "Ueber polarisierte Elektronenwellen," *Phys. Z.*, 31 (1930), 465–78.

67. F. Kirchner, "Ein Kathodenstrahl-Interferenzapparat für Demonstration und Strukturuntersuchungen," *Phys. Z.*, 31 (1930), 772–3.

68. G. P. Thomson, "Polarisation of Electrons," *Nature*, 126 (1930), 842.

69. O. Halpern, "Zur Reflexionspolarisation der Elektronenwellen," *Z. Phys.*, 67 (1931), 320–32.

70. E. Rupp, "Direkte Photographie der Ionisierung in Isolierstoffen," *Natur.*, *19* (1931), 109.

71. E. Rupp and L. Szilard, "Beeinflussung 'polarisierter' Elektronenstrahlen durch Magnetfelder," *Natur.*, *19* (1931), 422–3.

72. Mott (note 57).

73. E. G. Dymond, "Polarisation of a Beam of Electrons by Scattering," *Nature*, *128* (1931), 149.

74. Mott (note 57).

75. E. Rupp, "Versuche zum Nachweis einer Polarisation der Elektronen," *Phys. Z.*, *33* (1932), 158–64.

76. E. Rupp, "Neure Versuche zur Polarisation der Elektronen," *Phys. Z.*, *33* (1932), 937–40.

77. Mott (note 57).

78. E. Rupp, "Ueber die Polarisation der Elektronen bei zweimaliger 90°-Streuung," *Z. Phys.*, *79* (1932), 642–54.

79. N. F. Mott, "The Polarisation of Electrons by Double Scattering," *Proc. Roy. Soc. (London)*, *A135* (1932), 429–58.

80. Mott (note 79, p. 432).

81. J. Thibaud, J. J. Trillat, and T. von Hirsch, "Recherches sur la Polarisation d'un Faisceau d'electrons par Reflexion Cristalline," *Le Journal de Physique et le Radium*, *3* (1932), 314–19.

82. G. O. Langstroth, "Electron Polarisation," *Proc. Roy. Soc. (London)*, *A136* (1932), 558–68.

83. Langstroth (note 82, pp. 559–60).

84. Langstroth (note 82, pp. 566–7).

85. E. G. Dymond, "On the Polarisation of Electrons by Scattering," *Proc. Roy. Soc. (London)*, *A136* (1932), 638–51.

86. Dymond (note 85, p. 649).

87. Dymond (note 85, pp. 649–50).

88. Dymond (note 85, p. 650).

89. Dymond (note 85, p. 648).

90. Dymond (note 85, p. 639).

91. G. P. Thomson, "Polarisation of Electrons," *Nature*, *132* (1933), 1006.

92. E. Rupp, "Polarisation der Elektronen an freien Atomen," *Z. Phys.*, *88* (1934), 242–6. Professor H. R. Post has kindly pointed out to me an interesting aspect of Rupp's work. In 1935, Rupp published a note [*Z. Phys.*, *95* (1935), 801] in which he withdrew five papers because of mental illness that had allowed aspects of a dream world to intrude into his work. Rupp added that he saw no necessity to withdraw any earlier work. That was not, however, the view taken by Ramsauer [*Z. Phys.*, *96* (1936), 278], who stated that the group at Berlin had undertaken a critical review of Rupp's earlier work and was in fact repeating some of his earlier experiments on electron polarization. He also urged a note of caution in using Rupp's work without independent corroboration and suggested that Rupp's collaborators on some of the work might wish to reexamine their positions. No mention of this seems to appear in the subsequent literature. This revelation has no effect on my thesis that the problem of Mott scattering was the most important reason for the failure of scientists to reinvestigate

the experiments of Chase and Cox. Rupp's positive results only added to the confusion concerning the problem, and by the late 1930s, as we shall see later, the consensus in the physics community was that there was a discrepancy between Mott's theory and the experimental results. Professor Cox has informed me that during the 1930s he received a letter from an official in the Forschungs Institut der Allgemeinen Elektrischen Gesellschaft that indicated Rupp's difficulties. He supposes that it was sent to other workers in the field, and although he does not recall the author or its exact content, it seems probable that it was similar to the published letter of Ramsauer. Professor H. R. Post (private communication) was told that after Rupp's withdrawal, his locked laboratory was opened, and no apparatus for electron scattering was found. What was found was apparatus for forging data (H. Frauenfelder, private communication). Professor W. Paul, a student at Berlin at the time, also heard the same report (private communication).

93. F. Sauter, "Ueber den Mottschen Polarisationseffekt bei der Streuung von Elektronen an Atomen," *Annalen der Physik*, 18 (1933), 61–80.
94. E.G. Dymond, "On the Polarization of Electrons by Scattering. II," *Proc. Roy. Soc. (London)*, A145 (1934), 657–68.
95. Dymond (note 94, p. 662).
96. Dymond (note 94, p. 666).
97. G. P. Thomson, "Experiment on the Polarization of Electrons," *Phil. Mag.*, 17 (1934), 1058–71.
98. Thomson (note 97, p. 1062).
99. Thomson (note 97, pp. 1070–1).
100. Thomson (note 97, p. 1060).
101. E. Rupp, "Polarisation der Elektronen in magnetischen Feldern," *Z. Phys.*, 90 (1934), 166–76.
102. F. E. Myers, J. F. Byrne, and R. T. Cox, "Diffraction of Electrons as a Search for Polarization," *PR*, 46 (1934), 777–85.
103. This refers only to Dymond's 1932 report (note 85). A note added in proof indicated his 1934 retraction (note 94).
104. Myers et al. (note 102, p. 778).
105. H. Hellmann, "Bemerkung zur Polarisierung von Elektronenwellen durch Streuung," *Z. Phys.*, 96 (1935), 247–50.
106. O. Halpern and J. Schwinger, "On the Polarization of Electrons by Double Scattering," *PR*, 48 (1935), 109–10.
107. J. Winter, "Sur la polarisation des ondes de Dirac," *CR*, 202 (1936), 1265–6.
108. H. Richter, "Zweimalige Streuung schneller Elektronen," *Annalen der Physik*, 28 (1937), 533–54.
109. Richter (note 108, p. 554).
110. F. C. Champion, "The Single Scattering of Elementary Particles by Matter," *Reports on Progress in Physics*, 5 (1938), 348–60.
111. Champion (note 110, p. 358).
112. M. E. Rose and H. A. Bethe, "On the Absence of Polarization in Electron Scattering," *PR*, 55 (1939), 277–89.
113. Rose and Bethe (note 112, p. 278).

114. J. H. Bartlett and R. E. Watson, "The Elastic Scattering of Fast Electrons by Heavy Elements," *PR, 56* (1939), 612–13.
115. K. Kikuchi, "A Preliminary Report on the Polarization of Electrons," *Proceedings of the Physical-Mathematical Society of Japan, 21* (1939), 524–7. Kikuchi gave a more detailed report in "On the Polarization of Electrons," *Proceedings of the Physical-Mathematical Society of Japan, 22* (1940), 805–24.
116. M. E. Rose, "Scattering and Polarization of Electrons," *PR, 57* (1940), 280–8.
117. Rose (note 116, p. 280).
118. C. T. Chase and R. T. Cox, "The Scattering of 50-Kilovolt Electrons by Aluminum," *PR, 58* (1940), 243–51.
119. Chase and Cox (note 118, p. 243).
120. Chase and Cox (note 118, p. 248).
121. C. G. Shull, "Electron Polarization," *PR, 61* (1942), 198.
122. C. G. Shull, C. T. Chase, and F.E. Myers, "Electron Polarization," *PR, 63* (1943), 29–37.
123. G. Goertzel and R. T. Cox, "The Effect of Oblique Incidence on the Conditions for Single Scattering of Electrons by Thin Foils," *PR, 63* (1943), 37–40.
124. Shull et al. (note 122, pp. 34–6).
125. Bartlett and Watson (note 114).
126. Goertzel and Cox (note 123).
127. Shull et al. (note 122, p. 29).
128. A comprehensive review of the problem of electron polarization was published in 1956 by Tolhoek (note 46).
129. R. T. Cox, private communication.

CHAPTER 3

1. J. H. Christenson, J. W. Cronin, V. L. Fitch, and R. Turlay, "Evidence for the 2π Decay of the K_2^0 Meson," *PRL, 13* (1964), 138–40 ("the Princeton experiment").
2. G. D. Rochester and C. C. Butler, "Evidence for the Existence of New Unstable Elementary Particles," *Nature, 160* (1947), 855–7. For general background, see P. K. Kabir, *The CP Puzzle* (New York: Academic Press, 1978).
3. C. N. Yang, "Theoretical Interpretation of New Particles," in J. Ballam et al. (eds.), *High Energy Nuclear Physics*, proceedings of the sixth annual Rochester conference, 3–7 April 1956 (New York: Interscience, 1957), Chap. VIII, pp. 1–15.
4. A. Pais, "Some Remarks on the V-particles," *PR, 86* (1951), 663–72.
5. M. Gell-Mann, "Isotopic Spin and New Unstable Particles," *PR, 92* (1953), 833–4; T. Nakano and K. Nishijima, "Charge Independence for V Particles," *Progress in Theoretical Physics, 10* (1953), 581–2; K. Nishijima, "Some Remarks on the Even-Odd Rule," ibid., *12* (1954), 107–8; K. Nishijima, "Charge Independence Theory of V Particles," ibid., *13* (1955),

285–304; M. Gell-Mann, "The Interpretation of the New Particles as Displaced Charge Multiplets," *NC Suppl.*, *4* (1956), 848–66.

6. Yang (note 3, Chap. VIII, p. 4).

7. Baryon conservation, or conservation of particles with mass greater than or equal to the proton mass and with half-integral spin, has been proposed to explain the absence of otherwise allowed decays such as $p \rightarrow e^+ + \gamma$. No violations of this rule have been observed, although some theories of elementary particles now predict them. Stringent experimental tests are under way.

8. M. Gell-Mann and A. Pais, "Behavior of Neutral Particles Under Charge Conjugation," *PR*, *97* (1955), 1387–9. I shall consistently use the notation K_1^0, K_2^0 rather than the earlier θ_1^0, θ_2^0 or the later K_S^0, K_L^0.

9. Gell-Mann and Pais (note 8, p. 1389).

10. R. W. Thompson, "Decay Processes of Heavy Unstable Particles," *Progress in Cosmic Ray Physics*, *3* (1956), 255–337; W. H. Arnold, W. Martin, and H. W. Wyld, "Anomalous Neutral V-particles," *PR*, *100* (1955), 1545–7; J. Ballam, M. Grisaru, and S. B. Treiman, "Anomalous V-events as Three-body Decays," *PR*, *101* (1956), 1438–40; H. Blumenfeld et al., "Associated Production and the Lifetimes of the Λ^0 and θ^0 Particles," *PR*, *102* (1956), 1184–5.

11. K. Lande et al., "Observation of Long-lived Neutral V particles," *PR*, *103* (1956), 1901–4.

12. W. F. Fry, J. Schneps, and M. S. Swami, "Evidence for a Long-lived Neutral Unstable Particle," *NC*, *5* (1957), 130–43; K. Lande, L. M. Lederman, and W. Chinowsky, "Report on Long-lived K^0 Mesons," *PR*, *105* (1957), 1925–7; a similar report using a cloud chamber was J. A. Kadyk et al., "Cloud Chamber Investigation of Anomalous θ^0 Particles," *PR*, *105* (1957), 1862–71.

13. F. Eisler et al., "Systematics of Λ^0 and θ^0 Decay," *NC*, *5* (1957), 1700–15; "Associated Production of Σ^0 and θ_2^0, Mass of the Σ^0," *PR*, *110* (1958), 226–7.

14. A. Pais and O. Piccioni, "Note on the Decay and Absorption of the θ^0," *PR*, *100* (1955), 1487–9; cf. S. B. Treiman and R. G. Sachs, "Alternate Modes of Decay of Neutral K Mesons," *PR*, *103* (1956), 1545–9.

15. Francis Muller et al., "Regeneration and Mass Difference of Neutral K Mesons," *PRL*, *4* (1960), 418–21; M. L. Good, "Relation between Scattering and Absorption in the Pais-Piccioni Phenomenon," *PR*, *106* (1957), 591–5. If coherently regenerated (expected if K_1 and K_2 have very nearly the same mass), the K_1^0 would have the same momentum and direction as the K_2^0 that produced it.

16. I. B. Zel'dovich, "The Disintegration and Mass Difference of Heavy Neutral Mesons," *JETP*, *30* (1956), 1168–9; Treiman and Sachs (note 14). The empirical $\Delta S = \Delta Q$ rule states that in the leptonic decays of strange particles, the change in strangeness of the hadrons, the strongly interacting particles, must equal their change in charge. Thus, $\Sigma^- \rightarrow n e^- \nu$ is allowed, but $\Sigma^+ \rightarrow n e^+ \nu$ is forbidden. In the case of the K^0 mesons we can have only $\overline{K^0} \rightarrow \pi^+ e^- \nu$ and $K^0 \rightarrow \pi^- e^+ \nu$.

17. W. F. Fry and R. G. Sachs, "Method for Determining the θ_1–θ_2 Mass Difference," *PR*, *109* (1958), 2212–13.

18. Kabir (note 2, p. 12).
19. T. D. Lee, R. Oehme, and C. N. Yang, "Remarks on Possible Noninvariance Under Time Reversal and Charge Conjugation," *PR*, *106* (1957), 340–5.
20. L. Landau, "Conservation Laws in Weak Interactions," *JETP*, *32* (1957), 405–6.
21. H. H. Wyld, Jr., and S. B. Treiman, "Charge Asymmetries in the Decay of Long-lived Neutral K-mesons," *PR*, *106* (1957), 169–70.
22. M. Bardon et al., "Long-lived Neutral K mesons," *Annals of Physics*, *5* (1958), 156–81.
23. J. J. Sakurai, "$K_{\mu 3}$ Decay: Tests for Time Reversal and the Two Component Theory," *PR*, *109* (1958), 980–3.
24. S. Weinberg, "Time-reversal Invariance and θ_2^0 Decay," *PR*, *110* (1958), 782–4.
25. M. L. Good, "K_2^0 and the Equivalence Principle," *PR*, *121* (1961), 311–13.
26. Muller et al. (note 15); D. Neagu et al., "Decay Properties of K_2^0 Mesons," *PRL*, *6* (1961), 552–3.
27. R. G. Sachs, "Methods for Testing the CPT Theorem," *PR*, *129* (1963), 2280–5.
28. D. Luers et al., "K_2^0 Decay," *PR*, *133* (1964), B1276–89.
29. L. B. Leipuner et al., "Anomalous Regeneration of K_1^0 Mesons from K_2^0 Mesons," *PR*, *132* (1963), 2285–90.
30. J. W. Cronin, V. L. Fitch, R. Turlay, private communication.
31. V. Fitch, "The Discovery of Charge-Conjugation Parity Asymmetry," Nobel lecture, *Science*, *212* (1981), 989–93. The page of their laboratory notebook for 20 June 1963 is headed "CP invariance run."
32. J. H. Christenson, "Regeneration of K_1^0 mesons and the K_1^0–K_2^0 Mass Difference," PhD thesis, Princeton University, 1964, pp. 6, 18.
33. A. Abashian et al., "Search for CP Nonconservation in K_2^0 Decays," *PRL*, *13* (1964), 243–6.
34. Tran N. Truong, "Possibility of CP violation in $\Delta I = 3/2$ Decay of the K^0 Meson," *PRL*, *13* (1964), 358–61.
35. R. G. Sachs, "CP violation in K^0 decay," *PRL*, *13* (1964), 286–8; cf. Kabir (note 2).
36. The Princeton experiment had already offered evidence that what was being observed was $K_2^0 \hookrightarrow \pi^+ \pi^-$ in the calculated mass and direction of the decaying particle and the calculated mass of the decay particle.
37. K. Popper, *Conjectures and Refutations* (New York: Harper, 1968), 51.
38. Andrew Pickering, "The Role of Interests in High-Energy Physics; The Choice Between Charm and Colour," in Karin D. Knorr, Roger Krohn, and Richard Whitley (eds.), *The Social Progress of Scientific Investigation* (Dordrecht: D. Reidel, 1980), 107–38 (also in *Sociology of the Sciences*, *4*).
39. Roger H. Stuewer, *The Compton Effect* (New York: Science History Publications, 1975), particularly Chapter 1.
40. M. V. Terent'ev, "$K_2^0 \to 2\pi$ Decay and Possible CP-nonconservation," *Uspekhi Fizicheskikh Nauk*, *86* (1965), 231–62; cf. B. Laurent and M.

Roos, "On the Superposition Principle and CP Invariance in K⁰ Decay," *PL*, *13* (1964), 269–70.

41. J. J. Sakurai, *Invariance Principles and Elementary Particles* (Princeton University Press, 1964), 147–51.
42. Good (note 25).
43. Terent'ev (note 40); J. Prentki, "CP Violation," in *Oxford International Conference on Elementary Particles, 1965, Proceedings* (Oxford University Press, 1966), 47–58.
44. J. Bernstein, N. Cabibbo, and T. D. Lee, "CP Invariance and the 2π Decay of the K_2^0," *PL*, *12* (1964), 146–8.
45. Terent'ev (note 40).
46. Cf. Kabir (note 2, p. 20).
47. J. S. Bell and J. Perring, "2π decay of the K_2^0 Meson," *PRL*, *13* (1964), 348–9; Bernstein et al. (note 44).
48. Bell and Perring (note 47, p. 348).
49. N. Cabibbo, "Possibility of large CP and T Violation in Weak Interactions," *PL*, *12* (1964), 137–9; N. Cabibbo, "C-invariance Violation in Strong and Weak Interactions," *PRL*, *14* (1965), 965–8; T. D. Lee and L. Wolfenstein, "Analysis of CP-noninvariant Interactions and the K_1^0, K_2^0 system," *PR*, *138* (1965), B1490–6; J. Bernstein, G. Feinberg, and T. D. Lee, "Possible C, T Noninvariance in the Electromagnetic Interaction," *PR*, *139* (1965), B1650–9.
50. S. Weinberg, "Do Hyperphotons Exist?" *PRL*, *13* (1964), 495–7.
51. L. Lyuboshitz, E. O. Okonov, and M. I. Podgoretskii, "Galactic Hypercharge Field and the Two-pion Decay of the Longlived Neutral K Mesons," *Journal of Nuclear Physics*, *1* (1965), 490–6.
52. D. W. Sciama, "Mach's Principle and the Apparent Breakdown of CP Invariance in K Meson Dacays," *PL*, *13* (1964), 183.
53. T. D. Lee, "Hypercharge Conservation, CP Invariance and the Possible Existence of a Zero-mass Zero-spin Field," *PR*, *137* (1965), B1621–7.
54. G. Marx, "Decay $K_L^0 \rightarrow 2\pi$ and Spontaneous Breakdown of the CP Symmetry," *PRL*, *14* (1965), 334–6.
55. X. De Bouard et al., "Two Pion Decay of K_2^0 at 10 GeV/c," *PL*, *15* (1965), 58–61.
56. W. Galbraith et al., "Two-pion Decay of the K_2^0 Meson," *PRL*, *14* (1965), 383–6.
57. A. A. Grib, "Violation of CP Invariance in K⁰-Meson Decay and the Non-invariant Vacuum," *JETP Letters*, *2* (1965), 8–10.
58. Marx (note 54).
59. S. K. Kundu, "Electromagnetic 2π decay of K_2^0 – Meson Decays and the Non-invariant Vacuum," *Australian Journal of Physics*, *18*(1965), 395–400.
60. Richard Spitzer, "K_1^0 Regeneration by Magnetic Field and Apparent CP Violation," *Nuclear Physics*, *68* (1965), 691–4.
61. P. K. Kabir and R. R. Lewis, "Tests for Particle Mixture Theories of K⁰ $\rightarrow 2\pi$ Decay," *PRL*, *15* (1965), 306–8; P. K. Kabir and R. R. Lewis, "Further Discussion of Particle-mixture Theories of K⁰ $\rightarrow 2\pi$ Decay," *PRL*, *15* (1965), 711–14.

62. M. Levy and M. Nauenberg, "Apparent CP Violation Due to a New Vector Boson," *PL*, *12* (1964), 155–6.

63. N. Kalitzin, "The CP Invariance, the Paritino, and the Spurion," *Academie Bulgare des Sciences, Comptes Rendus*, *18* (1965), 627–9.

64. V. Fitch et al., "Evidence for Constructive Interference between Coherently Regenerated and CP-nonconserving Amplitudes," *PRL*, *15* (1965), 73–6.

65. G.B. Cvijanovich, E. A. Jeannet, and E. C. G. Sudarshan, "CP Invariance in Weak Interactions and the Pion Decay Asymmetry," *PRL*, *14* (1965), 117–18.

66. G. B. Cvijanovich and E. A. Jeannet, "Anisotropie dans la Desintegration π–μ Mesons π^+ Crees dans la Desintegration $K^+ \rightarrow 2\pi^+\pi^-$," *Helvetica Physica Acta*, *37* (1964), 211–12.

67. G. Rinaudo et al., "Existence of Pions with Spin," *PRL*, *14* (1965), 761–3.

68. S. Taylor et al., "Search for Anomalous π^+ Decay Among τ^+ Decay Secondaries," *PRL*, *14* (1965), 745–6.

69. H. J. Lipkin and A. Abashian, "A Possible Explanation for $K \rightarrow 2\pi$ Decays Without CP Violation," *PL*, *14* (1965), 151–3; J.L. Uretsky, "A Speculation Concerning the Apparent CP Violation in K^0 Decay," *PL*, *14* (1965), 154–6; H. Ezawa et al., "Particle Mixture Theory and Apparent CP Violation in K-Meson Decay," *PRL*, *14* (1965), 673–6; Kabir and Lewis (note 61).

70. K. Nishijima and M. J. Saffouri, "CP Invariance and the Shadow Universe," *PRL*, *14* (1965), 205–7.

71. A.E. Everett, "Evidence on the Existence of Shadow Pions in K^+ Decay," *PRL*, *14* (1965), 615–16.

72. A. Callahan and D. Cline, "Charged $K_{\pi 2}$ Branching Ratio," *PRL*, *15* (1965), 125–30.

73. B. P. Roe et al., "New Determination of the K^+-Decay Branching Ratios," *PRL*, *7* (1961), 346–8; F. S. Shaklee et al., "Branching ratios of the K^+ Meson," *PR*, *136* (1964), B1423–31.

74. L. A. Khalfin, "On the Quantum Theory of Unstable Elementary Particles," *Doklady Akademii Nauk SSSR*, *162* (1965), 1034–7; Terent'ev (note 40).

75. V. Fitch, C. A. Quarles, and H. C. Wilkins, "Study of the K^+ Decay Probability," *PR*, *140* (1965), B1088–91.

76. A. M. L. Messiah and O. W. Greenberg, "Symmetrization Postulate and Its Experimental Foundation," *PR*, *136* (1964), B248–67.

77. J. M. Gaillard et al., "Measurement of the Decay of the Long-Lived Neutral K Meson into Two Neutral Pions," *PRL*, *18* (1967), 20–5.

78. J. W. Cronin et al., "Measurement of the Decay Rate of $K_2^0 \rightarrow \pi^0 + \pi^0$," *PRL*, *18* (1967), 25–9.

79. Laurent and Roos (note 40); M. Roos, "On the Construction of a Unitary Matrix with Elements of Given Moduli," *Journal of Mathematical Physics*, *5* (1964), 1609–11.

80. A. Böhm et al., "Observation of Time Dependent K_S and K_L Interference

in the $\pi^+\pi^-$ Decay Channel from an Initial K^0 State," *PL*, *27B* (1968), 321–7, a preliminary report of which was given at the Heidelberg International Conference on Elementary Particles, 1967.

81. Articles by R. G. Sachs, "Phenomenologic Theory of CP-Violating K^0 Decays," in *High Energy Physics and Elementary Particles* (Vienna: 1965), 929–36; L. B. Okun, "Possible Properties of CP Violating Non-leptonic Interactions," ibid., 939–51; N. Byers, S. W. MacDowell, and C. N. Yang, "CP Violation in K-Decay," ibid., 953–80; T. D. Lee, "Weak Interactions and Questions of C, P, T Non-invariances," in *Oxford International Conference on Elementary Particles, 1965, Proceedings* (Oxford University Press, 1966), 225–39.

82. J. Steinberger, "CP Violation and K Decay," in A. Zichichi (ed.), *Recent Developments in Particle Symmetries* (New York: 1966), 205–47; J.S. Bell and J. Steinberger, "Weak Interactions of Kaons," in *Oxford International Conference on Elementary Particles, 1965, Proceedings* (Oxford University Press, 1966), 195–222.

83. Prentki (note 43); cf. Terent'ev (note 40).

84. U. Camerini et al., "Experimental Tests of Time-reversal Invariance in K Decay," *PRL*, *14* (1965), 989–91.

85. B. Aubert et al., "Leptonic Decays of K^0 Meson and PC Violation," *PL*, *17* (1965), 59–62.

86. M. Baldo-Ceolin et al., "Experimental Test of CP Invariance and $\Delta S = \Delta Q$ Selection Rule," *NC*, *38* (1965), 684–90.

87. P. Franzini et al., "Some Features of K^0 Decay," *PR*, *140* (1965), B127–34.

88. M. Roos, "Direct Tests of CP Invariance," *PL*, *20* (1966), 59–62.

89. Fitch (note 31, p. 990).

90. D. Hywel White, D. Sullivan, and E. J. Barboni, "The Interdependence of Theory and Experiment in Revolutionary Science: The Case of Parity Violation," *Social Studies of Science*, *9* (1979), 303–27.

91. Laurent and Roos (note 40, p. 269).

92. Bell and Perring (note 47, p. 348): "Clearly the theory has a very slender basis. However, it suggests a refinement of the experiment."

93. See S. Harding (ed.), *Can Theories Be Refuted?* (Dordrecht: D. Reidel, 1976).

94. T. P. Swetman, "The Response to Crisis – A Contemporary Case Study," *AJP*, *39* (1971), 1320–8 (on p. 1328); T.S. Kuhn, *The Structure of Scientifc Revolutions* (University of Chicago Press, 1962).

95. See Chapter 1.

96. K. Shrader-Frechette, "Atomism in Crisis; An Analysis of the Current High Energy Paradigm," *Philosophy of Science*, *44* (1977), 409–40, attempted to apply the idea of a Kuhnian crisis; R. E. Hendrick and A. Murphy, "Atomism and Illusion of Crisis; The Danger of Applying Kuhnian Categories to Current Particle Physics," ibid., *48* (1981), 454–68, persuasively refute the claim.

97. T. S. Kuhn, *Black-Body Theory and the Quantum Discontinuity, 1894–1912* (Oxford University Press, 1978).

98. T. S. Kuhn, "Revisiting Planck," *Historical Studies in the Physical Sciences*, *14(2)* (1984), 231–52.
99. D. Dorfan et al., "Charge Asymmetry in the Muonic Decay of the K_2^0," *PRL*, *19* (1967), 987–93.
100. Sheldon Bennet et al., "Measurement of the Charge Asymmetry in the Decay $K_2^0 \rightarrow \pi^{\pm} e^{\mp} \nu$," *PRL*, *19* (1967), 993–7.
101. R. C. Casella, "Time Reversal and the K^0 Meson Decays," *PRL*, *21* (1968), 1128–31; R. C. Casella, "Time Reversal and the K^0 Meson Decays. II.," *PRL*, *22* (1969), 554–6.
102. Casella (note 101 [1969], p. 555).

CHAPTER 4

1. There was another anomaly for Dirac theory involving the spectrum of hydrogen. See M. Morrison, "More on the Relationship Between Technically Good and Conceptually Important Experiments," *British Journal for the Philosophy of Science* (in press).
2. I. Hacking, *Representing and Intervening* (Cambridge University Press, 1983), Chapters 9–16. For a discussion of "good" experiments, defined on the basis of an experiment's relationship to existing theory or its call for new theories, see A. Franklin "What Makes a 'Good' Experiment?" *British Journal for the Philosophy of Science*, *32* (1981), 367–74. For an interesting discussion of experiment, see J. Worrall, "The Pressure of Light: The Strange Case of the Vacillating 'Crucial Experiment'," *Studies in History and Philosophy of Science*, *13(2)* (1982), 133–71.
3. T. S. Kuhn, *The Structure of Scientific Revolutions*, second ed. (University of Chicago Press, 1970), Chapter III.
4. A. Pickering, "Producing a World: Transformations of Experimental Practice in the History of High-Energy Physics" (in press).
5. See S. Harding (ed.), *Can Theories Be Refuted?* (Dordrecht: D. Reidel, 1976).
6. W. Quine, "Two Dogmas of Empiricism," in *From a Logical Point of View* (Cambridge: Harvard University Press, 1953), 43.
7. A. Pickering, "The Role of Interests in High-Energy Physics," *Sociology of the Sciences*, *IV* (1980), 107–38; A. Pickering, "The Hunting of the Quark," *Isis*, *72* (1981), 215–36.
8. Morrison (note 1).
9. L. Laudan has made an attempt to give weightings for various problems in *Progress and Its Problems* (Berkeley: University of California Press, 1977). I do not believe he has been successful.
10. C. Glymour, *Theory and Evidence* (Princeton University Press, 1980).
11. The current value is $1.232 \pm 0.024 \times 10^{-4}$, in good agreement with theory.
12. R. C. C. Leite, J. F. Scott, and T. C. Damen, "Multiple-Phonon Resonant Scattering in CdS," *PRL*, *22* (1969), 780–2.
13. J. F. Scott, private communication.
14. J. Dorling, "Bayesian Personalism, The Methodology of Scientific Research Programs, and Duhem's Problem," *Studies in History and Philos-*

ophy of Science, 10 (1979), 177–201; J. Dorling, "Further Illustrations of the Bayesian Solution of Duhem's Problem" (in press).

15. For comment and criticism of Dorling's work, see M. Redhead, "A Bayesian Reconstruction of the Methodology of Scientific Research Programs," *Studies in History and Philosophy of Science*, 11 (1980), 341–7.
16. Hacking (note 2, pp. 171–2); N. Hanson, *Patterns of Discovery* (Cambridge University Press, 1969).
17. B. Barnes, *T. S. Kuhn and Social Science* (New York: Macmillan, 1982), 65.
18. Kuhn offered the following statement on incommensurability: "These examples point to the third and most fundamental aspect of the incommensurability of competing paradigms. In a sense that I am unable to explicate further, the proponents of competing paradigms practice their trades in different worlds. . . . Equally it is why, before they can hope to communicate fully, one group or the other must experience the conversion that we have been calling a paradigm shift. Just because it is a transition between incommensurables, the transition between competing paradigms cannot be made a step at a time, forced by logic and neutral experience" (note 3, p. 150). "This need to change the meaning of established and familiar concepts is central to the revolutionary impact of Einstein's theory" (note 3, p. 102). Kuhn has recently modified his view. In an address to the Philosophy of Science Association (October 1982) he spoke of "local incommensurability," which I take to mean the difficulty of context-dependent translation. Feyerabend is even harder to pin down. At times he states that theories are incommensurable only if they are not interpreted in an "independent observation language" [*Against Method* (New York: New Left Books, 1975)]. Elsewhere he seems to deny the possibility of such a language: "Adopting the point of view of relativity we find that the experiments, *which of course will now be described in relativistic terms* [italics in original], are relevant to the theory, and we also find that they support the theory. Adopting classical mechanics we again find that the experiments *which are now described in the very different terms of classical physics* are relevant, but we also find that they undermine classical mechanics" (p.282).
19. The analysis also seems to apply to proton-proton or electron-electron scattering. An antirealist might object, however, that protons and electrons are theoretical entities whose definitions include mass. Thus, the two views of the experiments would be incommensurable, because they are talking about different entities. The advantage of billiard balls is that they can be defined ostensively: One can hold a billiard ball in one's hand or point to it.
20. Dr. Michael Redhead has graciously provided such a general proof: Let θ^* and ϕ^* be the scattering angles in the center of mass system (note: $\theta^* + \phi^* = 180°$), and let θ and ϕ be the angles in the laboratory system. Then $\tan\theta = \sqrt{2/\gamma+1}\ \tan \frac{1}{2}\theta^*$, where $\gamma = 1/\sqrt{1 - V_1^2/c^2}$, and similarly for ϕ. Then it can be shown quite easily that $\tan (\theta + \phi) = [2\sqrt{\gamma + 1}/(\gamma - 1)]/\sin \theta^*$, so that except for $\theta^* = 0°$, $\tan(\theta+\phi) < \infty$ or $\theta+\phi < 90°$.

21. P. Horwich, *Probability and Evidence* (Cambridge University Press, 1982), offers a good introduction to the subject.
22. See, for example, Dorling (note 14), Redhead (note 15), Horwich (note 21), and R. Rosenkrantz, *Inference, Method, and Decision* (Dordrecht: D. Reidel, 1977).
23. Horwich (note 21, Chapter 2).
24. C. Howson, "The Probability of Laws, and the Simplicity Postulate" (in press). Rosenkrantz (note 22) does offer an objective Bayesian view, but I am not convinced of its general applicability.
25. L. Savage, *The Foundations of Statistics*, second ed. (New York: Dover, 1982), 46–55, 68.
26. C. Howson, "Some Recent Objections to the Bayesian Theory of Support," *British Journal for the Philosophy of Science* (in press).
27. Howson (note 24); Rosenkrantz (note 22).
28. Much of the material in this section and the next was put together in collaboration with Dr. Colin Howson. I am grateful to him for allowing me to use it here.
29. This is similar to the Duhem–Quine problem discussed earlier. Instead of trying to decide which hypothesis is refuted by the evidence, here we try to decide how to allocate evidential support.
30. Howson (note 26); C. Howson, "Bayesianism and Support by Novel Facts," *British Journal for the Philosophy of Science*, *35(3)* (1984), 245–51. This paper provides a general Bayesian solution to the problem of ad hoc hypotheses.
31. D. Garber, "Old Evidence and Logical Omniscience in Bayesian Confirmation Theory," in J. Earman (ed.), *Testing Scientific Theories* (Minneapolis: University of Minnesota Press, 1983), 99–131.
32. I. Niiniluoto, "Novel Facts and Bayesianism," *British Journal for the Philosophy of Science*, *34* (1983), 375–9.
33. E. Zahar, "Why Did Einstein's Programme Supersede Lorentz's," *British Journal for the Philosophy of Science*, *24* (1973), 95. See also note 42.
34. J. Worrall, "The Ways in Which the Methodology of Scientific Research Programmes Improves on Popper's Methodology," in G. Radnitsky and G. Anderson (eds.), *Progress and Rationality in Science* (Dordrecht: D. Reidel, 1978), 45–70.
35. Bookmakers clearly make this distinction. They recognize the confirmation of the wager, but offer no credit, or cash, unless it was made prior to the event.
36. In 1728, Henry Pemberton wrote: "Now since each planet moves in an ellipsis, and the sun is placed in one focus; Sir Isaac Newton deduces from hence, that the strength of this power is reciprocally in the duplicate proportion of the distance from the sun," H. Pemberton, *A View of Sir Isaac Newton's Philosophy* (London: 1728), 172. A modern view is that "Newton justified his law of gravitation primarily by demonstrating that with it he could derive Kepler's laws of planetary motion," P. Tipler, *Physics* (New York: Worth, 1982), 349.
37. It is worthwhile to separate the discussion of gravitational force into two parts: (1) the dependence on the inverse square of the distance and (2) its

universal nature. The first, as shown in the text, follows straightforwardly from Kepler's Laws, whereas Newton's path to the second is quite complex. See R. S. Westfall, *Force in Newton's Physics*, (London: Macdonald, 1971), Chapter 7; C.A. Wilson, "From Kepler's Laws, So-Called, to Universal Gravitation: Empirical Factors," *Archive for the History of the Exact Sciences*, 6 (1970), 89–170. A Bayesian discussion of other aspects of this episode is contained in R. Laymon, "Newton's Demonstration of Universal Gravitation and Philosophical Theories of Confirmation," in Earman (note 31, pp. 179–99).

38. I. Newton, *Principia*, Motte's translation revised by Cajori (Berkeley: University of California Press, 1960).
39. All page numbers for citations from Newton refer to note 38.
40. As will be discussed later, this derivation applies to circular motion.
41. P. Duhem, *La Théorie Physique: Son Object, Sa Structure* (Paris: Marcel Riviere et Cie, 1906), Chapter VI; K. Popper, *Objective Knowledge* (Oxford University Press, 1972), Chapter V; E. Zahar, "The Popper-Lakatos Controversy," *Fundamenta Scientiae*, 3 (1982), 21–54.
42. J. Herivel, *The Background to Newton's Principia* (Oxford: Clarendon Press, 1965), 297, 301.
43. Westfall (note 37, p. 462).
44. A case in point was the construction of the Dubna 10-GeV synchrotron, which was a scaled-up version of the 3-GeV Brookhaven Cosmotron. The Dubna machine has never worked as well as it was supposed to, because certain difficulties that were negligible at 3 GeV became important at 10 GeV.
45. Horwich (note 21, pp. 118–19) has argued persuasively that diverse evidence provides more support for a hypothesis than narrow evidence by eliminating plausible competing hypotheses. His proof is as follows: Let $H_1, H_2, \ldots, H_k$ be exclusive, competing hypotheses, one of which is known to be true. By Bayes's Theorem,

$$P(H_1|E) = \frac{P(E|H_1)P(H_1)}{P(E)},$$

which can be written

$$P(H_1|E) = \frac{P(E|H_1)P(H_1)}{P(E|H_1)P(H_1) + P(E|H_2)P(H_2) + \ldots + P(E|H_k)P(H_k)},$$

Suppose H_1 entails both E_D and E_N where E_D and E_N are diverse and narrow evidence, respectively. Then

$$\frac{P(H_1|E_D)}{P(H_1|E_N)} = \frac{P(H_1) + P(E_N|H_2)P(H_2) + \ldots + P(E_N|H_k)P(H_k)}{P(H_1) + P(E_D|H_2)P(H_2) + \ldots + P(E_D|H_k)P(H_k)}$$

Because E_D is diverse evidence, there will be many cases H_j such that $P(H_j)$ is substantial and $P(E_D|H_j)$ is less than $P(E_N|H_j)$. Thus, $P(H_1|E_D) > P(H_1|E_N)$.

46. An example of such a collaboration is J. Lynch et al., "Peculiarities Observed in the Reaction $\pi p \rightarrow K\pi\Lambda$," *PL*, *35B* (1971), 457–60.

47. J.G. Williams et al., "New Test of the Equivalence Principle from Lunar Laser Ranging," *PRL*, *36* (1976), 555–8.

48. Committee on Gravitational Physics, Space Science Board, *Strategy for Space Research in Gravitational Physics in the 1980's* (Washington: National Academy Press, 1981), 44.

49. Williams et al. (note 47, p. 553).

50. S.H. Aronson et al., "Determination of the Fundamental Parameters of the K^0–$\overline{K}^0$ System in the Energy Range 30–110 GeV," *PRL*, *48* (1982), 1306–9.

51. Aronson et al. (note 50, p. 1309).

52. R. Messner et al., "New Measurement of the $K_L^0 \rightarrow \pi^+\pi^-$ Branching Ratio," *PRL*, *30* (1973), 876–9.

53. R. Piccioni et al., "Measurement of the Charge Asymmetry in the Decay $K_L^0 \rightarrow \pi^\pm\mu^\mp\nu$," *PRL*, *29* (1972), 1412–15.

54. Particle Data Group, "Review of Particle Properties," *Reviews of Modern Physics*, *52* (1980), S1–S286.

55. D. P. Stoker et al., "Search for Right-Handed Currents by Means of Muon Spin Rotation," *PRL*, *54* (1984), 1887–90; quotation from p. 1890. The earlier result was J. Carr et al., "Search for Right-Handed Currents in Muon Decay," *PRL*, *51* (1983), 627–30.

56. R. T. Cox, private communication. Professor Cox performed experiments during this period using both types of sources.

57. Particle Data Group, "Review of Particle Properties," *Review of Modern Physics*, *48* (1976), S1–S245; quotation from p. S81.

CHAPTER 5

1. P. Galison, "Theoretical Predispositions in Experimental Physics: Einstein and the Gyromagnetic Experiments, 1915–1925," *Historical Studies in the Physical Sciences*, *12(2)* (1982), 285–323; P. Galison, *How Experiments End* (in press).

2. M. Deutsch, "Evidence and Inference in Nuclear Research," *Daedalus* (Fall 1958), 88–98; quotation from pp. 97–8.

3. P. Galison, "How the First Neutral Current Experiments Ended," *Reviews of Modern Physics*, *55* (1983), 477–509; A. Pickering, "Against Putting the Phenomena First: The Discovery of the Weak Neutral Current," *Studies in History and Philosophy of Science*, *15(2)* (1984), 85–117.

4. F. Sciulli, "An Experimenter's History of Weak Neutral Currents," *Progress in Particle and Nuclear Physics*, *2* (1979), 41–87; quotation from p. 46.

5. UA1 Collaboration, "Experimental Observation of Events with Large Missing Transverse Energy Accompanied by a Jet of Photon(s) in $p\bar{p}$ Collisions at $\sqrt{s} = 540$ BeV," *PL*, *139B* (1984), 115–25.

6. G. Holton, "Subelectrons, Presuppositions, and the Millikan–Ehrenhaft Dispute," *Historical Studies in the Physical Sciences*, *9* (1978), 166–224.

7. "Millikan's aim was to prove that electricity really has the atomic structure, which, on the base of theoretical evidence, it was supposed to have.... By a brilliant method of investigation and by extraordinarily exact experimental technique Millikan reached his goal.... Even leaving out of consideration the fact that Millikan has proved by his researches that electricity consists of equal units, his exact evaluation of the unit has done physics an inestimable service, as it enables us to calculate with a higher degree of exactitude a large number of the most important physical constants." Presentation speech by A. Gullstrand, in *Nobel Lectures in Physics, 1922–41* (Amsterdam: Elsevier, 1965), 51–3.
8. R. A. Millikan, "On the Elementary Electrical Charge and the Avogadro Constant," *PR*, 2 (1913), 109–43.
9. Millikan had available a theoretical calculation against which to check his empirically determined parameter: E. Cunningham, "On the Velocity of Steady Fall of Spherical Particles Through a Fluid Medium," *Proc. Roy. Soc. (London), 83A* (1910), 357–65.
10. Using *m* and equation (5.1), we obtain

$$a^3 = \frac{3Fe_n v_g}{4\pi g(\sigma - \rho)(v_f + v_g)}$$

To complete the calculation, a value for *e* must be assigned. Both Millikan and I used $e = 4.78 \cdot 10^{-10}$.
11. "Since *n'* [our Δn] is always a small number and in some of the changes almost always had the value 1 or 2 its determination for any change is obviously never a matter of the slightest uncertainty. On the other hand, *n* is often a large number, but with the aid of the known values of *n'* it can always be found with absolute certainty as long as it does not exceed say 100 or 150." Millikan (note 8, pp. 123–4).
12. All notebook references are to folders 3.3 and 3.4 in the Millikan Collection. For information about the collection, see A.F. Gunns and J.R. Goodstein, *Guide to the Robert Andrews Millikan Collection* (New York: American Institute of Physics, 1975); J.R. Goodstein (ed.), *The Robert Andrews Millikan Collection, Guide to a Microfilm Edition* (Pasadena: California Institute of Technology, 1977).
13. The value of *g* is given on a sheet labeled "Density of clock oil taken by R. A. Millikan April 5, 1912," one of the undated pages at the end of the notebooks.
14. The distance *d* appears to the ½ power, because $\sqrt{v_g} = \sqrt{d/t_g}$. Millikan (note 8, p. 140) estimated the uncertainty in *e* due to *d* as 0.1%.
15. Two are given on an undated notebook page preceding the observation of 13 March 1912.
16. Notes for 28 October 1911 and 5 April 1912.
17. Cf. Millikan (note 8, pp. 112–15). In his final calculation of *e*, he took $\mu = 0.0001824$. I used $\mu = 0.0001825$ in all my calculations.
18. R. A. Millikan, "The Isolation of an Ion, a Precision Measurement of Its Charge, and the Correction of Stokes's Law," *PR, 32* (1911), 349–97; *Science, 32* (1910), 436–48.

19. F. Ehrenhaft, "Die Quanten der Elektrizität," *Annalen der Physik, 44* (1914), 657–700; F. Zerner, "Zur Kritik des Elementarquantums der Elektrizität," *Phys. Z. 16* (1915), 10–13; D. Konstantinowsky, "Elektrische Ladungen and brownsche Bewegung sehr kleiner Metallteilchen im Gase," *Annalen der Physik, 46* (1915), 261–87.
20. R. A. Millikan, "The Existence of a Subelectron?" *PR*, 8 (1916), 595–625.
21. Millikan (note 8, p. 139).
22. Millikan (note 8, p. 141).
23. On 19 December 1911, 27 January, 2 February, and 9 February 1912.
24. Dust gave low values of e because the density of oil plus dust is less than that of oil alone.
25. Millikan (note 8, p. 138).
26. Millikan excluded some values of t_f in calculating $D'(\Delta n,t)$: "In general too, only differences in $t_t - t_f$ amounting to as much as 20 seconds are used . . . since obviously the observational error is large when $t_t - t_f$ is small."
27. Millikan (note 8, p. 129).
28. W. M. Fairbank, Jr., and A. Franklin, "Did Millikan Observe Fractional Charges on Oil Drops?" *AJP, 50(5)* (1982), 394–7. I am grateful to Dr. Fairbank for allowing me to use our joint work here.
29. G. S. LaRue, J. D. Phillips, and W. M. Fairbank, "Observation of Fractional Charge of (1/3)e on Matter," *PRL, 46* (1981), 967–70, and the references cited in that paper.
30. R. A. Millikan, "A New Modification of the Cloud Method of Determining the Elementary Electrical Charge and the Most Probable Value of that Charge," *Phil. Mag., 19* (1910), 209–28.
31. "Such drops, both because of the smallness of their size and the smallness of their charge, are not in equilibrium with multiply charged drops and consequently evaporate so rapidly that their life is relatively short. The single observation mentioned above was probably upon such a drop, but it was evaporating so rapidly that I obtained a poor value e." Millikan (note 30, p. 223).
32. "The total number of changes which we have observed would be between one and two thousand and *in not one single instance has there been any change which did not represent the advent upon the drop of one definite invariable quantity of electricity or a very small multiple of that quantity*" (emphasis in original). Millikan (note 18). In his 1913 paper (note 8), which was devoted primarily to precise measurement of e, rather than to proving charge quantization, Millikan regarded his 1911 work as having definitively shown quantization.
33. Notebook references are to folders 3.3 and 3.4 in the Millikan Collection.
34. In addition, half the data on one drop were excluded by Millikan and by me because the voltage was fluctuating (note 8, p. 138).
35. For large values of $1/pa$, as Millikan himself suspected (note 8, p. 138), the first-order correction to Stokes's Law is insufficient.
36. "We get a very good estimate of n from the values of Δn, which is, in general, a very small number and easily obtained. Since n' [our Δn] is always a small number and in some of the changes almost always had the

value 1 or 2 its determination for any change is obviously never a matter of the slightest uncertainty. On the other hand, n is often a large number, but with the aid of the known values of n' it can always be found with absolute certainty as long as it does not exceed say 100 or 150." Millikan (note 8, pp. 123–4). As we shall see, Millikan was somewhat optimistic. The estimate of n becomes rather unreliable for values greater than 30.

37. This assumes that $v_g = $ (distance of fall)/t_g has already been calculated.

38. Thus, a drop with successive values of the total charge $n = 4, 3, 4, 5$ would be excluded, because there are only three unique values of the charge.

39. By comparison, Millikan omitted 12 observations on which he had performed calculations and 45 on which he did no calculations.

40. I tried to avoid exclusions except in cases in which a measurement was clearly inconsistent with the other values obtained for that drop. Including these measurements actually had no significant effect on our results. We note in passing that four or these excluded measurements came from the same anomalous drop (7 March 1912, fourth observation).

41. I have looked for the effects of evaporation by examining t_g, the time of fall, as a function of time, for each drop. No consistent evidence of such evaporation is seen.

42. The method used by Millikan in 1913 and in my modern reanalysis, whose primary purpose was to measure e precisely, did not look for the presence of fractional charge.

43. See note 32.

44. M. Gell-Mann, "A Schematic Model of Baryons and Mesons," *PL*, *8* (1964), 214–15; G. Zweig, "An SU$_3$ Model for Strong Interaction Symmetry and Its Breaking," CERN reports 8182/TH 401 (Jan. 1964) and 8419/TH/412 (Feb. 1964).

45. For an interesting history of the quark model, see A. Pickering, *Constructing Quarks* (University of Chicago Press, 1984). Although I think Pickering underestimates the role of experiment and overemphasizes the sociology of the physics community, he does present a substantial history.

46. L. W. Jones, "A Review of Quark Search Experiments," *Reviews of Modern Physics*, *49* (1977), 717–52, provides a survey up until 1977 and indicates failure of these searches. A very different and positive view is given by B. McCusker, *The Quest for Quarks* (Cambridge University Press, 1983). Professor McCusker's arguments seem unconvincing.

47. A very nice discussion of the controversy over these results, through 1977, is given in A. Pickering, "The Hunting of the Quark," *Isis*, *72* (1981), 216–36.

CHAPTER 6

1. Some writers on science have questioned this. They are, to use a statement by the contemporary British philosopher Gordon Sumner, "like blind men looking for a shadow of doubt." Mr. Sumner is better known as Sting, the lead singer of the rock group The Police. The quotation is from the song "King of Pain" from the album *Synchronicity*.

2. A preliminary discussion of this issue appears in A. Franklin, "The Epistemology of Experiment," *British Journal for the Philosophy of Science*, *35* (1984), 381–90.

3. I. Hacking, *Representing and Intervening* (Cambridge University Press, 1983).

4. D. Shapere, "The Concept of Observation in Science and Philosophy," *Philosophy of Science*, *49* (1982), 485–525.

5. Shapere (note 4, p. 492).

6. A. Franklin and C. Howson, "Why Do Scientists Prefer to Vary Their Experiments?" *Studies in History and Philosophy of Science, 15(1)* (1984), 51–62.

7. The theoretical context can also be important in determining "same" or "different." See note 6, pp. 54–5.

8. One should be careful here to distinguish between the theory of the apparatus and the theory of the phenomena or the theory under test.

9. Hacking (note 3, pp.194–7).

10. A Bayesian analysis shows this quite clearly. Let h be the statement "the apparatus is working properly" and let o be some observation such that $h \vdash o$. Then, if we observe o, $P(h \mid o) > P(h)$. It then follows that if the apparatus is working properly, we have more reason to believe in its results.

11. R. A. Millikan, "The Isolation of an Ion, a Precision Measurement of Its Charge, and the Correction of Stokes's Law," *PR*, *32* (1911), 349–97; quotation from p. 360.

12. R. A. Millikan, "A New Modification of the Cloud Method of Determining the Elementary Electrical Charge and the Most Probable Value of that Charge," *Phil. Mag.*, *19* (1910), 209–28; R. A. Millikan, "On the Elementary Electrical Charge and the Avogadro Constant," *PR*, *2* (1913), 109–43.

13. G. LaRue, J. Phillips, and W. M. Fairbank, "Observation of Fractional Charge of $(1/3)e$ on Matter," *PRL*, *46* (1981), 967–70.

14. LaRue et al. (note 13, p.967).

15. J. Phillips, "Residual Charge of Niobium Spheres," unpublished dissertation, Stanford University.

16. G. Arnison et al., UA1 Collaboration, "Experimental Observation of Isolated Large Transverse Energy Electrons with Associated Missing Energy at $\sqrt{s} = 540$ GeV," *PL*, *122B* (1983), 103–16; M. Banner et al., UA2 Collaboration, "Observation of Single Isolated Electrons of High Transverse Momentum in Events with Missing Transverse Energy at the CERN-$\bar{p}p$ Collider," *PL*, *122B* (1983), 476–85.

17. UA1 Collaboration (note 16, p. 104).

18. UA2 Collaboration (note 16, p. 477).

19. UA1 Collaboration (note 16, p. 115).

20. R. Sloanaker, "Apparent Temperature of Jupiter at a Wavelength of 10 cm," *Astronomical Journal*, *64* (1959), 346.

21. F. D. Drake and S. Hvatum, "Non-Thermal Microwave Radiation from Jupiter," *Astronomical Journal*, *64* (1959), 329–30.

22. Drake and Hvatum (note 21, p. 330).

23. G. Field, "The Source of Radiation from Jupiter at Decimeter Wavelengths," *Journal of Geophysical Research, 64* (1959), 1169–75.
24. Field (note 23, p. 1175).
25. G. Field, "The Source of Radiation from Jupiter at Decimeter Wavelengths. 2. Cyclotron Radiation by Trapped Electrons," *Journal of Geophysical Research, 65* (1960), 1661–71.
26. J. W. Warwick, "Theory of Jupiter's Decametric Radio Emission," *Annals of the New York Academy of Sciences, 95* (1961), 39–60.
27. J. W. Warwick et al., "Planetary Radio Astronomy Observations from Voyager 1 Near Saturn," *Science, 212* (1981), 239–43.
28. J. W. Warwick et al., "Planetary Radio Astronomy Observations from Voyager 2 Near Saturn," *Science, 215* (1982), 582–7.
29. J. W. Warwick, private communication.
30. Warwick et al. (note 27, p. 243).
31. Warwick et al. (note 28, p. 585).
32. Warwick et al. (note 28, p. 586).
33. Warwick et al. (note 28, p. 586).
34. J. H. Christenson, J. W. Cronin, V. L. Fitch, and R. Turlay, "Evidence for the 2π Decay of the K_2^0 Meson," *PRL, 13* (1964), 138–40.
35. Christenson et al. (note 34, p. 138).
36. Christenson et al. (note 34, p. 139).
37. It is interesting to note that various theoretical explanations to preserve CP symmetry suggested either that the decaying particle was not a K_2^0 meson or that the decaying particles were not pions. For a detailed history, see Chapter 3.
38. Unfortunately, the current trend toward publication of important results in letters journals sometimes precludes detailed discussion of these checks in the published paper, as Sir Brian Pippard noted in an address before the British Society for the Philosophy of Science, London, September 1980.
39. The probability quoted that the two distributions were the same was one in a million.
40. I was present at the talk.
41. D. F. Bartlett et al., "$n + p \rightarrow d + \gamma$ and Time Reversal Invariance," *PRL, 23* (1969), 893–7.
42. H. M. Randall, R. G. Fowler, N. Fuson, and J. R. Dangle, *Infrared Determination of Organic Structures* (New York: Van Nostrand, 1949).
43. Randall et al. (note 42, p. 97). A similar case occurred in a recent local trial. The defendant was convicted of heroin possession on the basis of infrared spectroscopy of a substance found in his possession. The spectrum of the substance matched that of heroin. An expert witness for the defense raised the legitimate objection that the spectroscope had not previously been run without the sample present to verify that there had been no prior contamination of the instrument. My own reaction on reading that in the newspaper was to note that if the instrument had been contaminated with heroin, and the trial substance was something other than heroin, then what would have been seen was the spectrum of the other substance superimposed on the heroin spectrum. After the trial was over, I spoke with a member of the jury and asked if my argument had been considered. He

told me that although the jury had considered the defense objection seriously, they opted for conviction because the defense had had ample opportunity to have the substance independently tested and had not done so. They inferred that the defense knew that the substance was heroin and avoided the test.

44. That was the reaction to a talk based on this work given at Chelsea College, London, in June 1984. Students often resort to the unknown source of error to explain discrepancies in their results. It is unusual to find this in professsional work.

45. Randall et al. (note 42, p. 97).

46. See discussion at the end of Chapter 1.

47. C. S. Wu et al., "Experimental Test of Parity Conservation in Beta Decay," *PR*, *105* (1957), 1413–15.

48. A more detailed examination of the strategies used in this experiment will be presented in the next chapter.

49. E. D. Commins and P. Kusch, "Upper limit to the Magnetic Moment of He^6," *PRL*, *1* (1958), 208–9.

50. F. R. Ottensmeyer et al., "Electron Microtephroscopy of Proteins," *Journal of Ultrastructure Research*, *52* (1975), 193–201; F.R. Ottensmeyer et al., "The Imaging of Atoms: Its Application to the Structure Determination of Biological Macromolecules," *Chemica Scripta*, *14* (1978–9), 257–62.

51. J. Dubochet, "Comment on Prof. Ottensmeyer," *Chemica Scripta*, *14* (1978–9), 293; A. Klug, "Direct Imaging of Atoms in Crystal and Molecules," *Chemica Scripta*, *14* (1978–9), 291–3.

52. Dubochet (note 51, p. 293).

53. This can be seen quite easily. Let us suppose, for simplicity, that there are only two competing explanations for the effect at m_0: (1) C, that the 3-S.D. effect is due to chance, and (2) T, a theory that predicts a particle at m_0. $P(C) + P(T) = 1 = P(T \mid e) + P(C \mid e)$, where e is the observation of the 3-S.D. effect at m_0. $P(T \mid e) - P(T) = P(C) - P(C \mid e)$, and because $P(T \mid e) > P(T)$, $P(C \mid e)$ is less than $P(C)$, and so the probability that the observation is due to chance goes down, increasing the validity of the observation.

54. J. Wolosewick and K. Porter, "Microtrabecular Lattice of the Cytoplasmic Ground Substance: Artifact or Reality?" *Journal of Cell Biology*, *82* (1979), 114–39; K. Porter and K. Anderson, "The Structure of the Cytoplasmic Matrix Preserved by Freeze-Drying and Freeze-Substitution," *European Journal of Cell Biology*, *29* (1982), 83–96.

55. See Porter and Anderson (note 54, p. 83) for a complete list of references. This includes not only work by Professor Porter and his collaborators but also work by others.

56. Porter and Anderson (note 54, p. 83).

57. Wolosewick and Porter (note 54, p. 114).

58. We note here that the argument is quite similar to that given for a changing source for the Saturn electric discharges discussed earlier. As long as the arguments establishing the existence of an object (i.e., the MTL) are still valid, then differences can be attributed to other sources.

59. Wolosewick and Porter (note 54, p. 135).
60. Wolosewick and Porter (note 54, p. 133).
61. Wolosewick and Porter (note 54, p. 133).
62. Wolosewick and Porter (note 54, p. 137).
63. Porter and Anderson (note 54, p. 83).
64. Porter and Anderson (note 54, p. 95).
65. Porter and Anderson (note 54, p. 83).
66. Further evidence in favor of the existence of the microtrabecular lattice, including some by methods other than electron microscopy, is given in K. R. Porter, "The Cytomatrix: A Short History of Its Study," *Journal of Cell Biology, 99* (1984), 3s–12s.
67. K. R. Porter, M. Beckerle, and M. McNiven, "The Cytoplasmic Matrix," in J. R. McIntosh (ed.), *Modern Cell Biology* (New York: Alan R. Liss, 1983), 259–302; quotation from p. 261.
68. Porter et al. (note 67, p. 290).
69. See, for example, H. Collins, "The Seven Sexes: A Study in the Sociology of a Phenomenon, or the Replication of Experiments in Physics," *Sociology, 6* (1976), 141–84; A. Pickering, "Against Putting the Phenomena First: The Discovery of the Weak Neutral Current," *Studies in History and Philosophy of Science, 15(2)* (1984), 85–117.

CHAPTER 7

1. I. Hacking, *Representing and Intervening* (Cambridge University Press, 1983), 186–209.
2. T. D. Lee and C. N. Yang, "Question of Parity Nonconservation in Weak Interactions," *PR, 104* (1956), 254–8; quotation from p. 255.
3. Lee and Yang (note 2, p. 257).
4. C. S. Wu et al., "Experimental Test of Parity Conservation in Beta Decay," *PR, 105* (1957), 1413–15.
5. Wu et al. (note 4, p. 1413).
6. E. Ambler et al., "Nuclear Polarization of Cobalt 60," *Phil. Mag., 44* (1953), 216–18.
7. Wu et al. (note 4, p. 1414).
8. Wu et al. (note 4, p. 1414).
9. R. L. Garwin, L. M. Lederman, and M. Weinrich, "Observations of the Failure of Conservation of Parity and Charge Conjugation in Meson Decays: The Magnetic Moment of The Free Muon," *PR, 105* (1957), 1415–17.
10. J. I. Friedman and V. L. Telegdi, "Nuclear Emulsion Evidence for Parity Nonconservation in the Decay Chain $\pi^+-\mu^+-e^+$," *PR, 105* (1957), 1681–2.
11. No other known particle has a lifetime of 2μsec.
12. Garwin et al. (note 9, p. 1415).
13. Garwin et al. (note 9, p. 1416).
14. The asymmetry observed in these different materials had different values but was clearly there. The differences were attributed to the different properties of the stopping materials.
15. Garwin et al. (note 9, p. 1416).

16. Friedman and Telegdi (note 10).
17. J. I. Friedman and V. L. Telegdi, "Nuclear Emulsion Evidence for Parity Nonconservation in the Decay Chain $\pi^+\to\mu^+\to e^+$." *PR, 106* (1957), 1290–3; Friedman and Telegdi (note 10).
18. Friedman and Telegdi (note 10, p. 1682).
19. Friedman and Telegdi (note 10, p. 1681).
20. The original asymmetry reported in the text was 0.062 ± 0.027. The value of 0.091 ± 0.022 was added in proof.
21. L. Landau, "Conservation Laws in Weak Interactions," *JETP, 32* (1957), 405–6.
22. J. H. Christenson et al., "Evidence for the 2π Decay of the K_2^0 Meson," *PRL, 13* (1964), 138–40.
23. Christenson et al. (note 22, p. 138).
24. Christenson et al. (note 22, p. 139).
25. Christenson et al. (note 22, p. 140).
26. For details, see Chapter 3.
27. For convenience, the experiment of Christenson et al. will be referred to as the Princeton experiment.
28. A. Abashian et al., "Search for CP Nonconservation in K_2^0 Decays," *PRL, 13* (1964), 243–6.
29. Abashian et al. (note 28, p. 244).
30. Abashian et al. (note 28, p. 244).
31. Abashian et al. (note 28, p. 245).
32. Abashian et al. (note 28, p. 245).
33. R. T. Cox, C. G. McIlwraith, and B. Kurrelmeyer, "Apparent Evidence of Polarization in a Beam of β-Rays," *Proceedings of the National Academy of Sciences (USA), 14* (1928), 544–9.
34. Cox et al. (note 33, p. 544).
35. Cox et al. (note 33, p. 546).
36. Cox et al. (note 33, p. 547).
37. Cox et al. (note 33, pp. 547–8).
38. C. Chase, "A Test for Polarization in a Beam of Electrons by Scattering," *PR, 34* (1929), 1069–74; quotation from p. 1070.
39. Chase (note 38, p. 1071).
40. Chase (note 38, p. 1072).
41. Chase (note 38, p. 1073).
42. C. Chase, "The Scattering of Fast Electrons by Metals. I. The Sensitivity of the Geiger Point-Discharge Counter," *PR, 36* (1930), 984–7.
43. Chase (note 42, pp. 986–7).
44. Chase (note 42, p. 987).
45. C. Chase, "The Scattering of Fast Electrons by Metals. II. Polarization by Double Scattering at Right Angles," *PR, 36* (1930), 1060–5; quotation from p. 1061.
46. Chase (note 45, pp. 1063–4).
47. L. Grodzins, "The History of Double Scattering of Electrons and Evidence for Polarization of Beta Rays," *Proceedings of the National Academy of Sciences (USA), 45* (1959), 399–405; L. Grodzins, "A Comment on the History of Double Scattering of Beta Rays," in B. Maglic (ed.), *Adventures*

in Experimental Physics, gamma volume (Princeton: World Science Education, 1973), 154–60.

48. See Chapter 2 for details.
49. G. Holton, "Subelectrons, Presuppositions, and the Millikan–Ehrenhaft Dispute," *Historical Studies in the Physical Sciences*, 9 (1978), 166–224.
50. R. A. Millikan. "A New Modification of the Cloud Method of Determining the Elementary Electrical Charge and the Most Probable Value of that Charge," *Phil. Mag.*, 19 (1910), 209–28.
51. Millikan (note 50, p. 212).
52. Millikan (note 50, pp. 213–4).
53. Millikan (note 50, p. 216).
54. Millikan (note 50, p. 219).
55. Millikan (note 50, p. 224).
56. Millikan (note 50, p. 220). Some commentators have suggested that this drop provides evidence in favor of fractionally charged free quarks. Millikan's explanation was that it was due to evaporation, and that has been generally accepted. In later work, Millikan did find an oil drop with charge 0.6e. That seems to have been simply an unreliable measurement. See discussion in Chapter 5.
57. Millikan (note 50, p. 224).
58. R. A. Millikan, "The Isolation of an Ion, A Precision Measurement of Its Charge, and the Correction of Stokes's Law," *PR*, 32 (1911), 349–97.
59. Millikan (note 58, p. 335).
60. Millikan (note 58, p. 360).
61. As we have seen, Millikan made a similar statement with respect to his 1913 results, and that was untrue. There is thus some reason to doubt this statement.
62. Millikan (note 58, pp. 382–3).
63. Millikan (note 58, p. 384).
64. R. A. Millikan, "On the Elementary Electrical Charge and the Avogadro Constant," *PR*, 2 (1913), 109–43.
65. Millikan (note 64, p. 120).
66. This is not true; see discussion in Chapter 5.
67. Millikan (note 64, p. 138). This statement is false. See discussion in Chapter 5.
68. Millikan's value of e given in his tables used a value of η, the viscosity of air, of 0.0001825. He later believed that the best value was 0.0001824 and used only that to calculate his final value of e.
69. Millikan (note 64, p. 141).
70. Millikan (note 64, p. 115).
71. Millikan (note 64, p. 115).
72. Millikan (note 58, p. 355).

CHAPTER 8

1. The accusations include the following: Summerlin's painting of black patches on the fur of white mice to simulate grafts [B.J. Culliton, "The

Sloan-Kettering Affair: An Uneasy Resolution," *Science*, *184* (1974), 1154–7]; Soman and Felig's fudged data on anorectic patients, which also included an accusation of plagiarism [M. Hunt, "A Fraud that Shook the World of Science," *New York Times Magazine* (1 November 1981), 42]; Long's establishment of cell lines, supposedly from human patients, that turned out to be owl monkey cells [N. Wade, "A Diversion of the Quest for Truth," *Science*, *211* (1981), 1022–5]; Straus's falsified patient data [W. J. Broad, "But Straus Defends Himself in Boston," *Science*, *212* (1981), 1367–9]; Gullis's invention of data concerning neuroblastoma and neuroblastoma/glioma hybrid cells ["Researcher Admits He Faked Journal Data," *Science News*, *111* (1977), 150–1]; Darsee's false data on drugs used to aid heart attack recovery [W. J. Broad, "Harvard Delays in Reporting Fraud," *Science*, *215* (1982), 478–82]; Spector's falsification of data concerning the difference between normal cells and cancer cells [S. Hughes, "Why Scientific Cheats Don't Get Away With It," *Denver Post* (21 February 1982), 3B]. Several of these cases, as well as some others, are discussed in William Broad and Nicholas Wade, *Betrayers of the Truth* (New York: Simon & Schuster, 1982). This work should, however, be used with some caution. In discussing Millikan's exclusion of data, they state that he "kept unfavorable results out of published papers while publicly maintaining that he had reported everything" (p. 227). Although it is true that Millikan did exclude data while claiming he had not, as was discussed in Chapter 5, only 1 of 107 excluded events can be regarded as unfavorable.

2. W. Schmaus, "Fraud and Sloppiness in Science," *Perspectives on the Profession*, *1* (1981), 1–4.
3. R. Merton, "The Normative Structure of Science," in *The Sociology of Science* (University of Chicago Press, 1973), 267–78.
4. Quoted in Schmaus (note 2).
5. C. Babbage, *Reflections on the Decline of Science in England* (London: 1830), 174–83.
6. N.F. Mott, "The Scattering of Fast Atomic Electrons by Nuclei," *Proc. Roy. Soc.* (*London*), *A124* (1929), 425–42.
7. N.F. Mott, "Polarisation of a Beam of Electrons by Scattering," *Nature*, *128* (1931), 454; N. F. Mott, "The Polarisation of Electrons by Double Scattering," *Proc. Roy. Soc.* (*London*), *A135* (1932), 429–58.
8. E. Rupp, "Mitteilung," *Z. Phys.*, *95* (1935), 801. De Broglie's comments are in the margin of a copy of Davisson's report to the International Electricity Congress of 1932, at the Ecole Normale Supérieure. O. Darrigol, "A History of the Question: Can Free Electrons Be Polarized?" *Historical Studies in the Physical Sciences*, *15(1)* (1984), 39–79; quotation from p.74.
9. In Tolhoek's review of the entire subject, "Electron Polarization, Theory and Experiment," *Reviews of Modern Physics*, *28* (1956), 277–98, Rupp's work is conspicuously absent.
10. R. A. Millikan, "On the Elementary Electric Charge and the Avogadro Constant," *PR*, *2* (1913), 109–43.
11. Babbage (note 5, p. 178).
12. Millikan (note 10, p. 138).

13. Some of the material in this section is based on work done in collaboration with Dr. Colin Howson. I am grateful to Dr. Howson for permission to use it here. He, of course, bears no responsibility for any errors.

14. J. H. Christenson, J. W. Cronin, V. L. Fitch, and R. Turlay, "Evidence for the 2π Decay of the K_2^0 Meson," *PRL*, *13* (1964), 138–40.

15. R. Messner et al., "New Measurement of the $K_L \to \pi^+\pi^-$ Branching Ratio," *PRL*, *30* (1973), 876–9; C. Geweniger et al., "A New Determination of the $K_2^0 \to \pi^+\pi^-$ Decay Parameters," *PL*, *48B* (1974), 487–91.

16. The confidence level for the mean of these four later measurements is 66%.

17. This might be considered the first observation of a grue quantity, or perhaps it is bleen.

18. J. Steinberger, private communication. I am grateful to Professor Steinberger for his comments and for providing me with a copy of the original proposal for the experiment, which reiterates his conclusions.

19. At the time, I was a member of the group performing the experiment. Although I was not an author of the paper, I did take part in the discussions concerning the measurement of $|\eta_{+-}|$.

20. Particle Data Group, "Review of Particle Properties," *Reviews of Modern Physics*, *48* (1976), S1–S245; quotation from p. S81.

21. S. H. Aronson et al., "Determination of the Fundamental Parameters of the K^0–$\overline{K}^0$ System in the Energy Range 30–110 GeV," *PRL*, *48* (1982), 1306–9.

22. D. P. Coupal et al., "Measurement of the Ratio $\Gamma(K_L \to \pi^+\pi^-)/\Gamma(K_L \to \pi l \nu)$ for K_L with 65 GeV/c Laboratory Momentum," *PRL*, *55* (1985), 566–9.

23. Particle Data Group, "Review of Particle Properties," *Reviews of Modern Physics*, *52* (1980), S1–S286; quotation from p. S286.

24. R. Birge, "A Survey of the Systematic Evaluation of the Universal Physical Constants," *NC* [*Suppl.*], *6* (1957), 39–67; quotation from p. 51.

25. E. R. Cohen and J. W. DuMond, "Our Knowledge of the Fundamental Constants of Physics and Chemistry in 1965," *Reviews of Modern Physics*, *37* (1965), 537–94; quotation from p. 551.

26. M. Henrion and B. Fischoff, "Uncertainty Assessment in the Estimation of Physical Constants" (in press).

27. Cohen and DuMond (note 25).

28. Cohen and DuMond (note 25, p. 551).

29. Cohen and DuMond (note 25, pp. 549–52); Birge (note 24, pp. 43–51).

30. Henrion and Fischoff (note 26).

31. Cohen and DuMond (note 25, pp. 551–2).

32. Cohen and DuMond (note 25, p. 541).

33. Geweinger et al. (note 15).

34. J. C. Cooper, "Have Faster-Than-Light Particles Already Been Detected?" *Foundations of Physics*, *9* (1979), 461–6. Cooper has also circulated two unpublished manuscripts on the same subject: "Fraudulent Experiment Won 1959 Nobel Prize," and "More on the Segrè Experiment."

35. O. Chamberlain, E. Segrè, C. Wiegand, and T. Ypsilantis, "Observations of Antiprotons," *PR*, *100* (1955), 947–50.

36. Cooper, "Segrè Experiment" (note 34).
37. Cooper's figures are actually 38.636364 ± 0.7342425 nsec, which claim a precision far in excess of that warranted by the data. They are, however, in good agreement with my own recalculation of 38.64 ± 0.73 nsec.
38. E. Segrè, private communication.
39. Cooper's figures are 51.116667 ± 1.4032851 nsec, again claiming far too much precision, but in good agreement with my value of 51.08 ± 1.40 nsec.
40. Background due to muons resulting from pion decay or electrons is negligible. Such contamination of particles traveling close to c would make the agreement better.
41. W.C. Harrison et al., "Reactions $\pi^{\pm}p \rightarrow pX^{\pm}$ (1100–1500 MeV) near Threshold," *PR*, *13D* (1976), 2453–68.
42. M. Kreisler, "Faster Than Light Particles – Do They Exist?" *Physics Teacher*, *13* (1975), 429–34.
43. Kreisler (note 42, p. 429).
44. This conclusion of the efficacy of repetition as a safeguard against fraud is in direct contradiction to that of Broad and Wade (note 1). They regard repetition as a figment invented by philosophers of science. The case histories demonstrate the incorrectness of their view.
45. Schmaus (note 2).
46. A similar view was expressed by H. Zuckerman, "Deviant Behavior and Social Control in Science," in E. Sagarin (ed.), *Deviance and Social Change* (Beverly Hills: Sage, 1977), 87–138; discussion on pp. 95–6. I was unaware of Professor Zuckerman's earlier work when I began this study.
47. Cooper, "Fraudulent Experiment" (note 34).
48. For a similar, but fictional, example, see the novel *The Affair* by C.P. Snow.
49. F. Bacon, *The New Organon and Related Writings*, F.H. Anderson (ed.) (New York: 1960).

Index

Abashian, A., 90, 205, 208
Abbe, Ernst, 167
absorbers, 84, 199, 213
accelerator beam, 84, 171; *see also* high-energy accelerators
AD, *see* average deviation
Adair, Robert K., 81, 82
ad hoc hypothesis, *see* experiment
Against Method (Feyerabend), 3
alcohol drops, 218
aldehydes, 185
Alihanov, 49
Allen, H.S., 217
Altman, Sidney, 49
aluminum, 61–2, 64, 68
Alvarez, Luis, 33, 170
Ambler, E., 17, 29
American Physical Society, 25, 27, 177
amino acids, 9
anisotropy, 194–6
antiproton, 239–41
"Apparent Evidence of Polarization in a Beam of β-Rays" (Cox et al.), 39, 71
Aristotelian mechanics (dynamics), 2, 21
Aronson, S.H., 130, 234, 235
Arsenieva, A.N., 54–5
associated production, 32, 73
asymmetry, 17, 18f, 19, 29–30, 42, 43–4, 45, 46, 47–8, 49, 50, 51, 53, 55–6, 57, 58, 60, 61, 62, 63, 65, 66, 68, 69, 70, 71, 79, 80, 126, 131, 194, 196, 197, 199, 201, 208, 210, 211, 214, 225
atomic spectra theory, 40
Aubert, B., 98
average deviation (AD), 161
Avogadro's number, 223

Babbage, Charles, 227, 230
Bacon, Francis, 243
Baldo-Ceolin, M., 98
Balmer series, 117
Barboni, Edward J., 26, 99, 100
Bardon, M., 80
Barkla, C.G., 45
Barnes, Barry, 110, 113
Bartlett, J.H., 67, 70
baryon conservation, 74, 258n7
Bayesian confirmation theory, 105, 108, 113–23, 124
Bayesianism, 114, 115
Bayes's Theorem, 114, 115, 120
Begeman, M., 217, 218
Bell, J.S., 93, 94,
Bennett, Sheldon, 102
Berkeley tables, 100
Bernstein, J., 93
beryllium, 57
beta particles, 43, 44, 45, 211
beta rays, 40, 42, 45, 49, 50, 55, 62, 65, 71
   asymmetry, 196, 197
   counter, 195
   decay, 14, 16, 17, 20, 23, 41, 47, 52, 62, 71, 72, 95, 135, 193–4, 208, 215
Bethe, Hans A., 67
Bevatron, 33
Birge, R., 236
Birman, J., 108
black-body radiation, 1, 113
Black-Body Theory and the Quantum Discontinuity 1894–1912 (Kuhn), 1
Bloch, Felix, 24
Block, Martin, 34
Bohr, Niels H.D., 54

Bohr theory, 104
Bose statistics, 98
bosons, 97, 101, 107
    W, 103, 140, 170–1, 190
Brancazio, Peter J., 25
Brookhaven National Laboratory,
    33, 81
Brownian movements, 220, 223
bubble chamber, 126, 129, 131, 132f,
    133, 139, 182
Butler, C.C., 31, 73

Cabibbo, N., 93
calibration, 175, 177, 190, 192, 205,
    222, 224
California Institute of Technology,
    142, 158
Callahan, A., 96
Camerini, U., 98
carbon, 18, 198, 199
Casella, R.C., 102
Cavanagh, P.E., 21
Cerenkov counter, 83, 84, 201, 202f
CERN, 139, 171
Chamberlain, O., 239
Champion, F.C., 66, 67
Chargaff, Erwin, 53
charge-conjugation invariance, 20,
    78, 80, 81, 82, 90, 91, 98, 100
    violation, 91, 92, 94–5, 103
charge conjugation parity combined
    symmetry (CP), 4, 30, 38, 74–5,
    78–80, 82, 87–90, 91, 92, 98, 102
    violation (1967), 73, 75, 79, 82, 89,
    90, 92, 93, 94, 95, 96, 97, 98, 99,
    102, 103, 106–7, 109, 125, 130,
    136, 175, 181, 193, 201–8, 225,
    232–8
    violation, acceptance of, 92–4, 97,
    98
charge quantization, 4, 140, 146, 158,
    162, 169, 215, 222
charge unit (e), 4, 147, 149, 150, 151,
    152, 153, 154, 157, 159–62, 163f,
    216, 220, 222, 223, 224
Chase, Carl T., 14, 39, 40, 41, 45–7,
    48, 49, 50, 51, 52, 54, 56, 57, 59,
    60, 62, 65, 66, 68, 70, 71, 72,
    211, 212–14, 215, 225
CHCl$_3$ *see* solvent
Christenson, J.H., 73, 90, 92, 100,
    201, 232

chronoscope, 142, 145, 154, 222
classical physics, 113, 127
Cline, D., 96
cloud chamber, 132t, 216, 217, 218
Co, *see* cobalt nuclei
cobalt (Co) nuclei, 17, 18f, 178, 181f,
    194, 197
CoCl$_2$ solution, 197
Cohen, E.R., 236, 237
collimators, 82f, 83, 202f
Columbia University, 243
    Physics Department, 24, 25
Committee on Gravitational Physics,
    129
Compton, A.H., 40
Compton scattering, 195
computer programs, 129–30
context of justification, 5
convection currents, 148, 158, 219,
    220, 222
cooking, 227, 232
Cooper, J.C., 239, 240, 241, 243
Copernican revolution, 101
Corske, Mark, 28
cosmic-ray physics, 104
cosmological model, 93, 98, 99
Cosmotron, 33
Coulomb scattering, 40, 41, 58, 61,
    66, 70
Coulomb's Law, 107
Coupal, D.P., 235
Cox, R.T., 14, 39, 40, 41, 43, 45, 46,
    47, 48, 49, 50, 51, 52, 53, 54, 55,
    56, 59, 60, 62, 65, 66, 68, 69, 70,
    71, 72, 208, 211, 212, 215, 225
CP, *see* charge conjugation parity
    combined symmetry
CPT theorem, 79, 80, 91–2, 98, 102,
    181
Cremonini, 168, 192
Cronin, J.W., 73, 81, 88, 90, 92, 97,
    102, 224, 232
Crowe, K.M., 26
Cs^{137} conversion line, 195
Cunningham, E., 219, 224
Curie, I., 213
Cvijanovich, G.B., 95, 97
cyclotron, 197, 200
cytoplasmic ground substance, 184,
    188

Dalitz, R.H., 12, 33

Dalitz plot, 12, 13
Darrow, Karl K., 25
Darsee, J., 243, 277–8n1
Darwin, C.G., 40
Davisson, C., 39, 40, 55, 208
De Bouard, X., 94
de Broglie, Louis, 39, 40, 53, 218, 228
decametric radiation, 173
deductive logic, 114–15
De Motu (Newton), 121
d'Espine, 213
deuterons, 240
Deutsch, Martin, 138
dextro (D)-rotary, 9, 10
Dirac, Paul A.M., 22, 55, 56, 58, 61, 63, 64, 65, 66, 67, 68, 72, 103, 105, 107, 108, 113, 115–16, 215, 225
Dorfan, D., 102
Dorling, John, 109
double scattering, *see* electron, scattering
double-slit interference experiment, 1, 2
Drake, F.D., 172
Dubochet, J., 180
Duhem, P., 3, 4, 36, 37, 100, 106, 119, 121
Duhem-Quine problem, 100, 103, 105–9
  Bayesian solution to, 109, 113, 117
DuMond, J.W., 236, 237
dust, 149, 174, 222, 270n24
Dymond, E.G., 58, 60–1, 62, 63, 64, 65, 67, 70

ebonite, 210
Ehrenhaft, Felix, 146, 147, 221, 231, 232
Einstein, Albert, 2, 110, 127, 241
Einsteinian mechanics, 110, 111
Eisler, F., 76
electric charge, 140, 146; *see also* charge quantization; charge unit
electric current, 9
electric-dipole, 11, 29–30
electric fields, 21, 216, 217, 219, 223
electromagnetic interactions, 15, 30, 140, 170; *see also* parity conservation
electron, 10, 12, 19, 20, 22, 83

accelerations, 54, 71, 72
  atomic, 63
  diffraction, 39, 40, 62
  free, 70
  micrograph, 178, 182–3f, 185, 187, 188f
  momentum, 16, 18, 41, 199
  polarization, 20–1, 40, 41, 42, 44–5, 46, 47, 48, 49, 51, 53, 54, 55, 56, 57–8, 59, 61, 62, 63, 64, 65, 66, 67, 68, 69–70, 71, 72, 135, 178, 211, 212, 224, 227
  scattering, 14, 21, 40–1, 43, 44, 45, 46, 48, 49, 50, 51, 54–5, 56, 57, 58, 59–60, 62, 64, 66, 67, 68–9, 70, 103, 108, 110, 135, 208, 209, 212, 214, 226; *see also* Rupp, E., fraudulent data and
  spin, 41–2, 46, 48, 49, 53, 54, 58, 66, 67, 208
  theory, 55, 56, 58, 61, 66–7, 68, 72, 107, 110, 227, 229; *see also* Dirac, Paul
  thermionic, 52, 62, 71, 72, 135
  velocities, 46, 48, 51, 56, 62, 63, 213, 215
  wave vector, 39
electron microscopy, 167, 168, 180, 186, 190
electroscope, 46, 57, 213, 214
electrostatic deflector, 21
elementary particle physics, 22, 24, 29, 30, 75
ellipses, 119, 121, 123
emulsions, 131, 132t
energy, conservation of, 111, 112
Eötvös experiments, 93, 94
ether, 2, 113
evaporation, 217, 218
event-fitting program (SQUAW), 129
Everett, A.E., 96
excitons, 108
experiment
  ad hoc hypothesis and, 100, 103
  apparatus, 3, 33, 35, 37, 44, 45, 46, 48, 50–1, 54, 60–1, 62, 63, 67, 71, 72, 82–4, 86–7, 107, 109–10, 126–7, 129, 130, 131–3, 135, 141, 142, 148, 149, 152, 158, 165, 167–8, 169, 170, 175, 176, 177–8, 179, 190, 192, 195–7, 199, 201, 202f, 210–11, 212, 213–14,

experiment (*cont.*)
215, 219, 221f, 222, 224, 225, 272n8, 273n43
bandwagon effect, 233, 235, 238, 241, 242
convincing, 103
crucial, 4, 7, 21–9, 35, 52, 103, 109, 193
different, 129–37, 166
experimenter and, 138, 140, 167, 192, 236
extrapolation and, 168
neglect of, 1, 8, 244
repetition of, 127, 129, 166, 229, 237–8, 242
results of, 1, 2, 3–4, 5–6, 7, 35–6, 39, 43–4, 46–8, 57, 61, 65, 100, 123–4, 127, 128, 130, 138, 139–40, 165, 166, 170, 214t, 215, 220–1, 222, 223–4, 226, 241, 242
selected analysis in, 147–54, 226, 227, 229–30, 231, 242, 271n39
theory and, 3, 4–5, 7–8, 52, 54, 63, 65–8, 71, 72, 99–100, 101, 103–5, 108, 109–10, 127, 139, 165, 166–7, 170, 173, 241, 244, 272n8
validation, 166, 167, 170, 171, 175, 181, 185, 186–7, 190–1, 192–3, 215, 218, 225
see also charge-conjugation invariance; charge-conjugation parity combined symmetry; fraud; oil-drop experiments; parity nonconservation
exponential law, 96, 101, 107

Fairbank, W.M., 162, 163, 164, 169, 170, 238
Falsification, 4
faster than light particles, *see* tachyons
Felig, P., 243, 277–8n1
Fermi interactions, 15
fermions, 97
Fermi's theory, 22
Feyerabend, Paul, 3, 110
Feynman, R.P., 24, 34, 100
Field, G., 172, 173
Fischoff, B., 237
Fitch, V.L., 73, 81, 90, 95, 96, 98, 99, 102, 224, 232
Flamsteed, John, 122
Fletcher, Harvey, 219

Forging, 227
fractional charge, 157–8, 159, 161, 162–3, 169
Franken, Peter, 236
Franzini, P., 98
fraud, 157, 226, 227–43
Frauenfelder, Hans, 11, 20, 24
freeze-drying, 185, 188–9f
Frenkel, J., 55, 60
Fresnel, Augustin Jean, 2
Friedman, Jerome L., 8, 18, 20, 29, 197, 199, 200
Frisch beam, 83, 84
Fry, W.F., 77
Fues, E., 57

Gaillard, J.M., 97
Galbraith, W., 94
Galileo, 1–2, 168, 192
Galison, Peter, 3, 138, 139
gamma decay, 33
gamma-radiation, 55
gamma ray, 24, 25, 33, 43, 45, 194, 210, 212, 213, 214, 224
anisotrophy, 196
counter, 195
Ganguly, A., 108
Garber, D., 118
Garwin, Richard L., 8, 18, 26, 27, 28, 29, 52, 103, 197, 200, 224
Geiger, Hans, 218
Geiger counter, 43, 44, 46, 48, 50, 224
Geiger tubes, 210, 213
Gell-Mann, M., 32, 34, 74, 75, 76, 77, 79, 80, 92, 99, 100, 163
Germer, L.H., 39, 40, 55, 208
gestrichene, see stroked state
Geweniger, C., 238
Gibson, W.M., 9
glutaraldehyde, 187, 188f
glycerine, 220, 222
Glymour, Clark, 3, 4, 108
Goertzel, G., 69, 70
gold, 43, 49, 55, 57, 58, 61, 62, 64, 67, 213
Golley, 226
Good, M.L., 77, 80, 92
Gooding, David, 3
Goudsmit, Samuel, 27, 28, 39, 40, 53, 181, 208
gravitational interactions, 37

gravitational mass, 80
Greenberg, O.W., 97
Grib, A.A., 94
Grodzins, Lee, 40, 42, 49, 50
Gullis, R., 243, 277–8n1
gyromagnetic experiments, 138, 139

Hacking, Ian, 3, 4, 103, 165, 166, 167, 192
hadrons, 139, 140, 163
Halliday, David, 24
Halpern, O., 58, 60, 66
Handschy, Mark, 2–3
Harmonices Mundi (Kepler), 169
Hayward, R.W., 17, 29
Heisenberg, Werner, 53
helium, 82, 87, 205
^{6}nuclei, 179, 190
Hellman, H., 57, 66
Henley, Ernest M., 11, 24
Henrion, M., 237
Heraclitean repetitions, 127, 129
Herschel, William, 121
high-energy accelerators, 33, 104
high-energy physics, 104, 182
Holton, Gerald, 138, 140, 141, 150, 215
Hoppes, D.D., 17, 29
Howson, Colin, 118, 119
Hudson, R.P., 17, 29
Hvatum, S., 172
hydrogen, 108, 205, 208; *see also* tritons
hyperon decay, 14, 15, 31, 32, 74
hyperphoton, 93

incommensurability, *see* theory-ladenness of observation and incommensurability
inertial mass, 80
"intellectual phase locking," 236
intrinsic parity, 10, 11, 30
Introductory Nuclear Physics (Halliday), 24
iron atoms, 10–11
isomers, 9, 247n9

Jeannet, E.A., 95
Joffé, A.F., 54
Jovanovic (neutron monitor), 83, 84
Jupiter, 104, 120, 121, 122, 168, 172, 173, 190, 192

Kabir, P.K., 95
Kalitzin, N., 95
kaons, *see* mesons, K
Kepler, Johannes, 120, 122, 169
Kepler's Laws, 119, 120, 121, 123
Third, 169, 192
Kikuchi, K., 67, 68
Kirchner, F., 57, 61
Klein, O.B., 22
Konstantinowsky, D., 147
K particles, 11, 12, 13, 31, 73, 74–81; *see also* meson, K
Kreisler, Michael, 241
Kuhn, Thomas S., 1, 3–4, 101, 110, 113
Kuhnian crisis, 100–1
Kundu, S.K., 95
Kurlbaum, F., 1, 113
Kurrelmeyer, B., 39, 45, 50, 51, 52, 56, 65, 71, 208, 212

laboratory notebooks, 83–7, 141, 142–3, 144f, 154, 158, 205
laevo (L)-rotary, 9, 10
Lakatos, I., 4
Landau, L., 24, 79, 80, 91, 107, 201
Lande, K., 75, 76
Langstroth, G.O., 60, 64
Laporte, Otto, 10, 11
Laporte's Rule, 11
Laurent, B., 97, 100
Lawrence, E.O., 236
lead, 45, 49, 84, 130
Leaning Tower of Pisa, 1
least squares fit (LSF), 161
Lederman, Leon M., 18, 25, 28, 29, 101, 197
Lee, T.D., 8, 12, 13, 14, 15, 16, 17, 18, 20, 22, 23, 24, 25, 26, 27, 28, 30, 31, 33, 34, 35, 36, 37, 50, 52, 53, 72, 78, 79, 93, 94, 193, 199
left-hand rule, 9, 22, 25, 43
left-right asymmetry, 21, 133, 135t
leptonic decay, 81
Levy, M., 95, 98
Lewis, R.R., 95
light, 2, 40, 108
linear drift, 2–3
Lipkin, H.J., 21
liquid hydrogen, 82
Long, J., 243, 277–8n1
longitudinal polarization, 20, 21, 41, 42, 47, 49, 62, 71, 72, 135, 199

LSF, *see* least squares fit
Lüders, G.C., 79
Lummer, O., 1. 113
Lyuboshitz, L., 94

McIlwraith, C.G., 39, 45, 50, 52, 65, 71, 208, 212
magnet, 83, 84, 201, 202f
magnetic cloud chamber, 75
magnetic field, 9, 19, 28, 58, 64, 83, 95, 133, 173, 179, 180, 196, 198, 200, 201, 211, 213
   crossed, 21
manometer, 216
Mars, 120
Marshak, R., 34, 100
Marshall, Leona, 23
Marx, G., 94
Massachusetts Institute of Technology (MIT), 84
Mercury, 120
mercury, 220, 222
   thermometer, 110
Merton, R., 226
meson, 10, 11, 12, 13, 15, 20, 33, 90, 131, 132t, 135t, 194, 199, 240
   beam, 132, 197, 199
   decay, 14, 15, 18–19, 20, 27, 75, 76, 77, 79, 80, 81, 82, 87–8, 89, 90, 92, 94, 95–6, 97, 98, 99, 100, 102, 130, 201, 205, 208
   K, 31, 32, 73, 74, 75, 76–7, 78, 79, 80–1, 82, 87, 91, 92, 93, 94, 95, 96, 97, 98, 99, 100, 102, 107, 130, 136, 175–6, 190, 201–8, 224
Messiah, A.M.L., 97
Michel, G., 34, 35
Michelson, A., *see* Michelson-Morley experiment
Michelson-Morley experiment, 1, 2
microscope, 166, 168, 192
microtrabecular lattice (MTL), 166, 184–9, 191
microtubules, 167
microwave radiation, *see* sychrotron radiation
Millikan, Robert A., 4, 140, 169, 190; *see also* oil-drop experiments
Millikan Collection (California Institute of Technology), 142, 158

mirror symmetry violation, *see* parity nonconservation
MIT, *see* Massachusetts Institute of Technology
Molecular Beams (Ramsey), 30
momentum conservation, 79, 111, 112
Monte Carlo calculation, 87, 88f, 177, 202. 203f
moon, 121
Moreau, J.J., 218
Morley, E., *see* Michelson-Morley experiment
Motion, Third Law of, 121
Mott, Nevill F., 40, 41, 46, 53, 55–6, 57, 58, 59, 60, 61, 63, 64, 65, 66, 67, 68, 69, 71, 72, 116, 225, 227
Mott scattering, 21, 40–1, 42, 47, 48, 54, 59, 61, 66, 67, 70, 71, 103, 107, 215
MTL, *see* microtrabecular lattice
Muller, Francis, 77, 81, 92
muon, 19, 20, 80, 83, 131, 132, 134t, 139, 197, 198, 199–200, 201, 205, 206, 207
   decay, 18, 133
   filter, 130
   spin rotation, 133, 194, 200
muonium, 200
Myers, F.E., 65
myokinase, 180, 181, 182–3f

Nauenberg, M., 95, 98
Neagu, D., 81
Neptune, 121
neutrino, 95, 139, 165
neurtron, 139, 205
   intensity, 83, 84
Nevis cyclotron, 197
Newton, Isaac, 119, 120, 121–3
Newtonian mechanics (dynamics), 21, 109, 110, 111, 127, 128
Newton's Inverse-Square Law of Universal Gravitation, 119–20, 121, 122–3
nickel crystals, 55
Niiniluoto, I., 118, 119
niobium spheres, 158, 170
Nishijima, K., 32, 74, 96, 99
Nobel Prize, 23, 24, 25, 239
   1957, 22, 25, 53
   1980, 102

nuclear emulsions, 20, 199
nuclear physics, 24
nuclei, 16, 17, 18, 29; *see also* Mott
  scattering
nucleons, 32, 73

observation, *see* theory-ladenness of
  observation and
  incommensurability
Oehme, R., 20, 79
oil-drop experiments, 4, 140–62, 193,
  215–24, 225, 226, 268n7,
  fraud and, 229–32
oil spectrum, 177–8, 190
Okonov, E.O., 94
"On the Absence of Polarization in
  Electron Scattering" (Rose and
  Bethe), 67
"On the Elementary Electrical
  Charge and the Avogadro Con-
  stant" (Millikan), 229
Oppenheimer, J.R., 35
Orear, J., 33, 34
osminium tetroxide, 185, 186
Ottensmeyer, F.R., 180
Oxford Conference on Elementary
  Particles (1965), 97–8

Pais, A., 32, 73, 74, 75, 76, 77, 79,
  80, 92, 99
paritino, 95
parity concept, 8–9, 10, 11, 12, 24,
  34–5
parity conservation, 8, 10–11, 13, 14–
  15, 16–17, 23, 24–5, 28, 34, 37,
  38, 80, 193, 247n7
  violation, *see* parity
    nonconservation
parity nonconservation (1957), 4, 7,
  8, 11–38, 41, 51, 73, 79, 80, 91,
  101, 103, 106, 107, 109, 110,
  193–201, 208–15
  acceptance of, 25–9, 35, 90
  attitudes toward, 23–24, 52–72
  experiments (1928, 1930), 7, 8, 14,
    20, 39–40, 43–9, 50–1, 71–2, 208,
    212–14
particle-antiparticle invariance, *see*
  charge-conjugation invariance
particle-antiparticle symmetry, *see*
  charge conjugation parity combined symmetry

particle detection techniques, 33
Paul, W., 228
Pauli, Wolfgang, 23–4, 25
Perrin, J., 218
Perring, J., 93
photoelectrons, 43, 209, 210, 212
photon, 11, 140
Physical Review, 27, 181
Piccioni, O., 76, 92, 99
Pickering, Andrew, 3, 104, 138, 139
pion, 10, 11, 12, 13, 32, 73, 74, 75,
  76, 80, 83, 87, 101, 102, 133t,
  139, 198, 199, 201, 205, 239,
  240, 241
  decay, 75, 77, 80, 81, 82, 94, 95,
    97, 98, 107, 206
  shadow, 96
Planck, Max, 1, 113, 218, 223
platinum points, 43, 210, 213
Podgoretskii, M.I., 94
Pollard, B.R., 9
Popper, Karl R., 3, 91, 119, 121
Porter, Keith, 184, 189
positron, 108, 115, 133, 200
positronium, 200
Post, H.R., 228
Prentki, J., 92, 97
Princeton experiment, 81–90, 92, 94,
  98, 99, 103, 175–7, 205, 208
Principia (Newton), 119–20, 122
Pringsheim, E., 1, 113
protamine, 180, 181, 183f
proton, 31, 132, 134t, 171, 240
Przibram, Karl, 147, 221
psychasthenia, 228
PtK$_2$ cell, 187, 188f
Purcell, E.M., 29, 30

quantization, 1, 113, 140, 146, 158,
  162; *see also* charge quantization
quantum-mechanics, 9, 10, 22, 24,
  53, 59, 101, 104, 113, 117, 118,
  126
quarks, 155, 163, 171
"Question of Parity Conservation in
  Weak Interactions" (Lee and
  Yang), 14
Quine, W., 3, 4, 36, 37, 106; *see also*
  Duheme-Quine problem

Rabi, I.I., 25
radiation, 10, 11

radioastronomy, 167, 168, 170
radiotelescope, 168, 190
radium, 43, 45, 55, 210, 212, 213
Ramsey, N.F., 24, 29, 30
reflection experiment, 69–70
Reflections on the Decline of Science in England (Babbage), 227
reflection symmetry violation, *see* parity nonconservation
Regener, Victor H., 218, 220, 223
relativistic electron theory, *see* Dirac, Paul A.M.
relativity, theory of, 2, 22, 92, 101, 110, 127, 241
Representing and Intervening (Hacking), 3
"Review of Particle Properties" (Particle Data Group), 131, 136, 137, 234, 236, 238
Richter, H., 66, 67, 70
Riehl, N., 44, 211
right-hand rule, 9, 22, 25, 43
Rinaudo, G., 95
Rochester, G.D., 31, 73
Rochester conference on high energy nuclear physics (1956), 32, 33, 34
Roos, M., 97, 98, 100
Rose, M.E., 67
Rose-Gorter method, 194
Rosenfeld, Arthur, 182
Rubbia, C., 97
Rubens, H., 1, 113
Rupp, E., 54, 57, 58, 59, 61, 62, 63, 64, 65, 68, 226
    fraudulent data and, 227–9
Rutherford, F., 217, 218

Sachs, R.G., 77, 81
Saffouri, M.J., 96
Sakurai, J.J., 80
Salam, A., 24, 140, 170, 171
Saturn, 120, 121, 122
    electric discharges (SED), 174–5, 190, 191
    magnetosphere, 174, 175
Sauter, F., 63
Savage, L., 114, 115
"Scattering of Fast Electrons by Metals, The. II. Polarization by Double Scattering at Right Angles" (Chase), 39

Schmaus, W., 242
Schrödinger, Erwin, 40
Schwinger, J., 66
scientific change, 3, 101
scintillation counters, 83, 131, 132, 133t, 201, 202f
Sciulli, Frank, 139
Scott, James, 108
S.D., *see* standard deviation
SED, *see* Saturn, electric discharges
Segrè, E., 239, 240, 241, 243
semileptonic decays, 77, 98
Shapere, Dudley, 3, 4, 165
Shapiro, A.M., 31
Shapiro, M.M., 31
Shull, C.G., 69
Siegbahn, K., 14
Sloanaker, R., 172
sodium iodide counters, 195
solar system, 121, 123
solvent ($CHCl_3$), 177, 179f, 190
Soman, V., 243, 277–8n1
Space inversion, 8, 9, 10
spark chambers, 83, 86, 126, 131, 132t, 133, 201, 202f
spatial parity, 10
special relativity, 126, 127, 128, 241
spectrometers, 82–3, 130, 131, 132t, 201
spectroscopy, 52, 53, 104
    atomic, 108
    infrared, 177, 178–9f, 190, 273n43
spin, 11, 12, 13, 16, 18, 19, 21, 23, 34, 40, 92, 93, 95, 97, 179, 180, 198, 200; *see also* electron
spion, 95, 96
Spitzer, Richard, 95
SPS, *see* Super Proton Synchrotron Collider
SQUAW, *see* event-fitting program
standard deviation (S.D.), 27, 28, 48, 87, 89, 161, 176, 181, 182, 184, 204, 233, 238
Standard Model, 140
Starry Messenger (Galileo), 168
Steinberger, Jack, 233, 234
Stern-Gerlach apparatus, 179, 190
Stokes's Law, 141, 145, 147, 150, 152, 158, 160, 217, 219, 221, 222, 223, 224, 230
strangeness conservation, 32, 38, 99

strange particles, 15, 31, 32, 33, 34, 73, 74, 93
stroked state, 10, 11
strong interactions, *see* parity conservation
Structure of Scientific Revolutions, The (Kuhn), 3
Stuewer, Roger H., 3
Sudarshan, E.C.G., 95, 100
Sullivan, Daniel, 26, 99, 100
sun, 121, 123
superconductivity, 241
superposition principle, 97, 98, 107
Super Proton Synchrotron (SPS) Collider, 171
Swetman, T.P., 100–1
synchrotron radiation, 104, 172–3, 190
Szilard, L., 58, 65

tachyons, 239, 240, 241
"tacking paradox," 117
Taylor, S., 95
TCP theorem, *see* CPT theorem
Telegdi, V.L., 8, 18, 20, 23, 25, 26, 27, 28, 29, 52, 101, 103, 181, 197, 199, 200
telescopes, 83, 169, 170, 198, 201
    Galilean, 168
temperature, 17, 18f, 110, 142, 145, 148, 156, 172, 178, 216
Terent'ev, M.V., 92
thallium vapor, 62
theory, 1, 2, 3, 118; *see also* experiment
theory-ladenness of observation and incommensurability, 105, 109–13
theta-tau puzzle, 8, 11–12, 13, 14, 15, 23, 31, 33, 34, 35, 52, 54, 78, 193; *see also* parity nonconservation
Thibaud, J., 59
Thomson, G.P., 57, 61, 62, 63–4, 65
Thomson, J.J., 216
three-view geometric program (TVGP), 129
time reversal, 79, 80, 91, 98, 102, 176–7
Tolhoek, H.A., 53
transmission experiment, 69, 70
transverse polarization, 20, 21, 49, 55
Treiman, S.B., 77, 80

trimming, 227, 230, 231t, 232, 242
tritons, 240
tungsten, 60, 87, 175, 176, 203, 205
Turlay, R., 73, 81, 90, 232
TVGP, *see* three-view geometric program

Uhlenbeck, G.E., 39, 40, 208
ungestrichene, see unstroked state
University of California, 182
unstroked state, 10, 11
Uranus, 121

V-A theory of weak interactions, 26, 36, 80, 100, 104, 108
vacuum, 94, 95
Van Allen belts, 172
vector interactions, 108
velocity, 2, 110, 111, 113, 127–8, *see also* electron, velocities
Venus, 120
voltmeter, 154, 222, 223
Voyager 1 and 2, 174, 175

Warwick, J.W., 173
water drop observations, 216, 218
Watson, R.E., 67, 70
wave function, 10, 12, 24, 97
wave theory, 2
weak interactions, 13, 14, 15, 16, 20, 24, 26, 32, 35, 36–7, 38, 39, 47, 51, 74, 75, 78, 80, 95, 96, 98, 99, 101, 110, 140, 193
weak neutral currents, 139
Weinberg, S., 80, 93, 140, 170, 171
Weinrich, Marcel, 18, 29, 197
Weisskopf, Victor, 24, 25
Weyl, H. 23
Wheaton, Bruce, 3
White, D. Hywel, 26, 99, 100
Wick, G.C., 30
Wien's Law deviations, 1
Wightman, A.S., 34
Wigner, Eugene, 11, 22, 34
Wilson, H.A., 216, 217
Winter, J., 66
Wolf, F., 54
Worrall, John, 2, 3, 118, 119
Wu, C.S., 8, 14, 17, 23, 24, 25, 26, 27, 28, 29, 52, 101, 103, 106, 178, 194–7, 224
Wyld, H.H., 80

x-ray, 45, 91

Yale University, 243
Yang, C.N., 8, 11, 13, 14, 15, 16, 17,
    18, 20, 22, 23, 24, 25, 27, 30, 31,
    32, 33, 34, 35, 36, 37, 50, 52, 53,
    72, 73, 78, 79, 193, 199
Young, Thomas, 1, 2

Yuan, Chia-Liu, 23
Yukawa particle, 241

Zahar, E., 118, 119, 121
Zel'dovich, I.B., 77
Zerner, F., 147
Zumino, B., 79
Zweig, G., 163